Renewable Energy - Recent Advances

Edited by Ahmed M.A. Nahhas and Akaehomen O. Akii Ibhadode

Published in London, United Kingdom

Renewable Energy - Recent Advances
http://dx.doi.org/10.5772/intechopen.100760
Edited by Ahmed M.A. Nahhas and Akaehomen O. Akii Ibhadode

Contributors
Akawu Shekari Biliyok, Salawudeen Ahmed Tijani, Geon Hwa Ryu, Young Gon Kim, Chae Joo Moon, Ji Ye Park, Ah Reum Lee, I. Made Wartana, Ni Putu Agustini, Samuel A. Alagbada, Umesh C. Rathore, Sanjeev Singh, Fabien Chidanand Robert, Jun Peng, Xiang Li, Ejiroghene Thelma Akhihiero, Akaehomen O. Akii Akii Ibhadode, Osezua Ibhadode, Ahmed M.A. Nahhas

First published in London, United Kingdom, 2023 by IntechOpen
IntechOpen is the global imprint of INTECHOPEN LIMITED, registered in England and Wales, registration number: 11086078, 5 Princes Gate Court, London, SW7 2QJ, United Kingdom

British Library Cataloguing-in-Publication Data
A catalogue record for this book is available from the British Library

Additional hard and PDF copies can be obtained from orders@intechopen.com

Renewable Energy - Recent Advances
Edited by Ahmed M.A. Nahhas and Akaehomen O. Akii Ibhadode
p. cm.
Print ISBN 978-1-80356-527-9
Online ISBN 978-1-80356-528-6
eBook (PDF) ISBN 978-1-80356-529-3

For EU product safety concerns:
IN TECH d.o.o., Prolaz Marije Krucifikse Kozulić 3, 51000 Rijeka, Croatia,
info@intechopen.com or visit our website at intechopen.com.

Meet the editor

Prof. Dr. Ahmed M.A. Nahhas received a Ph.D. in Electrical Engineering (Electronics) from the University of Pittsburgh, USA, in 2001. He served in many prestigious leadership positions, including the dean of the College of Engineering, department head, and vice dean. He is currently a full professor in the Department of Electrical Engineering, the College of Engineering and Islamic Architecture, Umm Al-Qura University, Saudi Arabia. He has reviewed several academic articles and dissertations in electrical engineering, electronics, and communications engineering. His current research interests include developing new photonic and electronic devices at nanoscales involving various functional materials, such as wide-bandgap semiconductors, nanostructured materials, and fabrication of ZnO and GaN optical devices. Prof. Dr. Nahhas received his first patent in 2016 and is an official reviewer of several international journals.

Akii Ibhadode is a Distinguished Professor of Manufacturing Engineering and former Shell Professor of Lightweight Automobile Engine Development. Previously, he was the vice-chancellor of the Federal University of Petroleum Resources, Nigeria. He has a BSc, MEng, and PhD in Mechanical Engineering. He has pioneered research leading to patents and industrial products. He was the winner of the Nigeria Prize for Science in 2010 and the winner of the Edwin Walker Prize of the Institution of Mechanical Engineers, UK, in 1988. He has supervised more than 100 MSc and Ph.D. students. He is a fellow of many professional societies and editor-in-chief of the *International Journal of Engineering Research in Africa* (JERA) published by Trans Tech Publications, Switzerland, which he founded in 2009.

Contents

Preface

This book provides an overview of recent progress in renewable energy materials and devices. Various forms of renewable energy, such as solar, water, and wind, have garnered significant research attention due to their potential applications. These sources have emerged as primary contenders for various uses. When compared to traditional energy sources, renewable energy offers numerous advantages and possibilities. The book comprises 11 chapters, each focusing on distinct aspects. The introductory chapter sets the stage, followed by chapters two to four which delve into devices. Chapters five and six concentrate on materials, while chapters seven and eight explore biofuels. Lastly, chapters nine through eleven examine microgrids.

Ahmed Nahhas
Department of Electrical Engineering,
College of Engineering and Islamic Architecture,
Umm Al Qura University,
Makkah, Saudi Arabia

Akaehomen O. Akii Ibhadode
Department of Production Engineering,
University of Benin,
Benin City, Nigeria

Section 1

Introduction

Chapter 1

Introductory Chapter: Overview of Renewable Energy

Ahmed M.A. Nahhas

1. Introduction

This chapter presents an introductory review about renewable energies. Renewable energy is the most important resource of reproducible infinite energy [1]. There are major differences between renewable and regular energy resources. These differences include the diversity, availability, and potential everywhere. Moreover, renewable energy produces less carbon dioxide [2–4]. Renewable energy has several forms. These forms include solar, water, wind, geothermal, and biomass. Wind energy is produced from air and wind. It is produced due to the heat of the Earth's surface. The more the wind speed increases, the more energy is produced. On the other hand, the solar energy is produced from the sunlight. Special silicon cells are used to collect sunlight. The resulting collected energy is stored on batteries. Several types of batteries are used for storing solar energy. Biomass energy is another form of renewable energy. This type of energy is produced from biomaterials. Geothermal energy is produced from heat inside the earth [5, 6].

2. Renewable energy technologies

There are several technologies used for producing the renewable energy. These technologies include electrical and mechanical to produce heat and electricity. These technologies are used for hydro, solar, fuels, and geothermal energy [7]. Renewable energy resources are divided into several categories. These categories differ from country to country and from land to land. Different countries have different policies for the renewable energy categories. The political perspective of the policy maker in one country may be to justify different treatment for established resources such as large hydroelectric from nascent resources such as geothermal [8].

3. Different types of renewable energy sources

There are several types of renewable energy sources including:

1. Geothermal resources. This type of energy requires drilling in the ground to be extracted. Drilling is very expensive and requires a lot of time.

2. Hydroelectric resources. This type of energy is extracted from the seashore and water resources. It mainly depends on the water movements.

3. Wind energy. Wind energy depends on the wind speed and direction. It is constructed on the shores, mountains, and flat areas.

4. Solar energy. This type of energy is extracted from the sun. Some countries like Saudi Arabia, UEA, and other Arab countries are very suitable for the solar energy.

5. Biomass energy. This type of energy is extracted from wood and waste, agricultural waste and residue, and crops. The cost of this type of energy is very high.

4. Renewable energy applications

Renewable energy has several applications. These applications can be divided based on two categories: ON Grid and the OFF Grid. The ON Grid includes solar and wind energy [8, 9]. The OFF Grid includes biomass and geothermal energies. The most important factor for these two categories is the cost of connections and installations. The OFF Grid is used for homes and small businesses. On the other hand, ON Grid is used for cities and big factories. Moreover, a third category, mini Grid, is used for mixed applications [9].

5. Benefits of renewable energy

The advantages of renewable energy are numerous and affect the economy, environment, national security, and human health. Some benefits include reliability, security, and resilience of power grid, job creation industries, and reduced carbon emissions and air pollution [9–11].

Author details

Ahmed M.A. Nahhas
Faculty of Engineering and Islamic Architecture, Department of Electrical Engineering, Umm Al Qura University, Makkah, Saudi Arabia

*Address all correspondence to: amnahhas@uqu.edu.sa

References

[1] Bhatia S, Gupta R. Renewable Energy. UK: WPI; 2018

[2] Sorensen B. Renewable Energy. Elsevier; 2017

[3] Twidell J, Weir T. Renewable Energy Resources. Taylor and Francis; 2015

[4] Alabdali Q, Bajawi A, Nahhas A. Review of recent advances of shading effect on PV solar cells generation. Sustainable Energy. 2020;**8**(1)

[5] Alzahrani G, Alzahrani F, Nahhas A. Study of the specific factors effecting the PV solar cell's efficiency in Saudi Arabia. Sustainable Energy. 2020;**8**(1)

[6] Alabdali Q, Bajawi A, Fatani A, Nahhas A. Review of recent advances in wind energy. Sustainable Energy. 2020;**8**(1)

[7] Alzahrani G, Nahhas A. Analytical study of power generation by photovoltaic system for Al-Riyadh and Al-Jubail regions in Saudi Arabia. American Journal of Energy Research. 2021;**9**(1)

[8] Alabdali Q, Nahhas A. Simulation study of grid connected photovoltaic system using PVsyst software: Analytical study for Yanbu and Rabigh regions in Saudi Arabia. American Journal of Energy Research. 2021;**9**(1)

[9] Bajawi A, Nahhas A. Analytical study of power generation using PV system for Al-Shuaiba and Al-Shuqiq regions in Saudi Arabia. American Journal of Energy Research. 2021;**9**(1)

[10] Albarrak A, Alshareef A, Alshareef A, Nahhas A. Review of hydrogen based fuel cells energy storage systems. Sustainable Energy. 2022;**10**(1)

[11] Jee A, Alsluimani J, Alqurashi A, Akkur A, Nahhas A. Review of recent advances of supercapacitors energy storage systems. Sustainable Energy. 2022;**10**(1)

Section 2

Renewable Energy Devices

Chapter 2

Structural Design Strategies for the Production of Internal Combustion Engine Components by Additive Manufacturing: A Case Study of a Connecting Rod

Osezua Ibhadode

Abstract

Topology optimization and lattice design strategies are excellent tools within the design for additive manufacturing (DfAM) workflow as they generate structurally optimal, lightweight, and complex features often difficult to produce by conventional manufacturing methods. Moreover, topology optimization approaches are quickly evolving to accommodate AM-related processes and geometric constraints. In this study, the re-design of the connecting rod of an internal combustion engine (ICE) is explored by topology optimization and lattice structures. In both topology optimization and lattice design, the objective is to maximize their structural performances while constraining material usage. Structural analyses are carried out on the optimized topologies to compare their mechanical performances with a benchmark design. Results show that the redesign of the connecting rod through topology optimization alone can realize 20% material savings with only a 5% reduction in the factor of safety. However, the combination of topology optimization and lattice structure design can result in over 50% material savings with a 21–26% reduction in the factor of safety. For manufacturability, the fast-predictive inherent strain model shows the designs through topology optimization and lattice design gives the lowest process-induced deformations before and after support structure removal.

Keywords: topology optimization, design for additive manufacturing, lattice structures, internal combustion engine, connecting rod

1. Introduction

There is a projection that about 80% of powertrains will still depend on carbon-based fuels by the year 2050 [1] despite the advancements in other fuel and energy sources such as hydrogen, biofuels, solar, batteries, etc. To reduce the adverse impact of carbon-based fuels on the environment and to comply with the Fourth Industrial Revolution, it is imperative to reduce CO_2 emission levels to the barest minimum and deploy autonomous and advanced manufacturing processes. Two of the several ways to meet up with the objectives for ICEs are the use of lightweight materials and additive manufacturing. Considering lightweight materials, alloys of aluminum, titanium, magnesium, and other rare earth metals [2, 3] are great for ICEs however they come at high prices [4]. Some efforts have yielded the development of hybrid aluminum composites with inexpensive locally sourced materials made from palm kernel shell (PKS) and periwinkle shell (PS) [5].

Additive manufacturing (AM) has been identified as the manufacturing technology for Industry 4.0 by the World Manufacturing Forum for several reasons not limited to the potential savings in material and energy costs, the elimination of tooling, process flexibility, and the ability to produce intricate features. Considering this, deliberate efforts have gone into designing and redesigning parts suitable for AM. The rapid technological developments of additive manufacturing (AM) processes due to intensive research efforts have sustained their capabilities to produce functional end-use parts. Being a layer-by-layer production process, the realization of complex structural features is now possible, enabling the exploration of new design strategies for optimal structures. Design for additive manufacturing frameworks have been developed, and they enable the user or designer to effectively explore new design pathways. Over the years, research and industry have shown that topology optimization and lattice design form the most popular structural design strategies for additive manufacturing. While design for additive manufacturing is presented in Section 2, topology optimization and lattice designs are presented in Section 3. AM process simulation is presented in Section 4 while the original and optimized designs are compared for functionality, in the static condition, and manufacturability.

2. Design for additive manufacturing

In 2007, Rosen [6] succinctly defined Design for Additive Manufacturing (DfAM) as the "synthesis of shapes, sizes, geometric mesostructures, and material compositions and microstructures to best utilize manufacturing process capabilities to achieve desired performance and other life-cycle objectives". There are four key aspects of the definition: multiscale structure/geometry, material, manufacturing process, and performance. To put it simply, design for additive manufacturing utilizes multiscale structures (micro, meso, and macro) and materials to achieve the desired performance while considering the additive manufacturing process. DfAM essentially sets itself apart from other design-for-manufacturing frameworks due to its ability to apply multiscale structural design approaches. Several DfAM frameworks have been introduced in the past decade, and several of these frameworks are summarized in [7]. Tang et al. [8] proposed an integrated topological and functional optimization framework, Yang and Zhao [9] suggested an additive manufacturing-enabled design framework, Primo et al. [10] developed a product design framework for AM with the integration of topology optimization, Orquera et al. [11] proposed a multifunctional optimization methodology for DfAM while Lei et al. [12] implemented an AM process model for product family design. While there are several other proposed frameworks, one thing common to most models is the adoption of an optimization tool. Taking a critical look into the models will reveal that topology optimization and lattice design form the basis of structural design methodologies. The AM-enabled design framework developed by Yang and Zhao is shown in **Figure 1**.

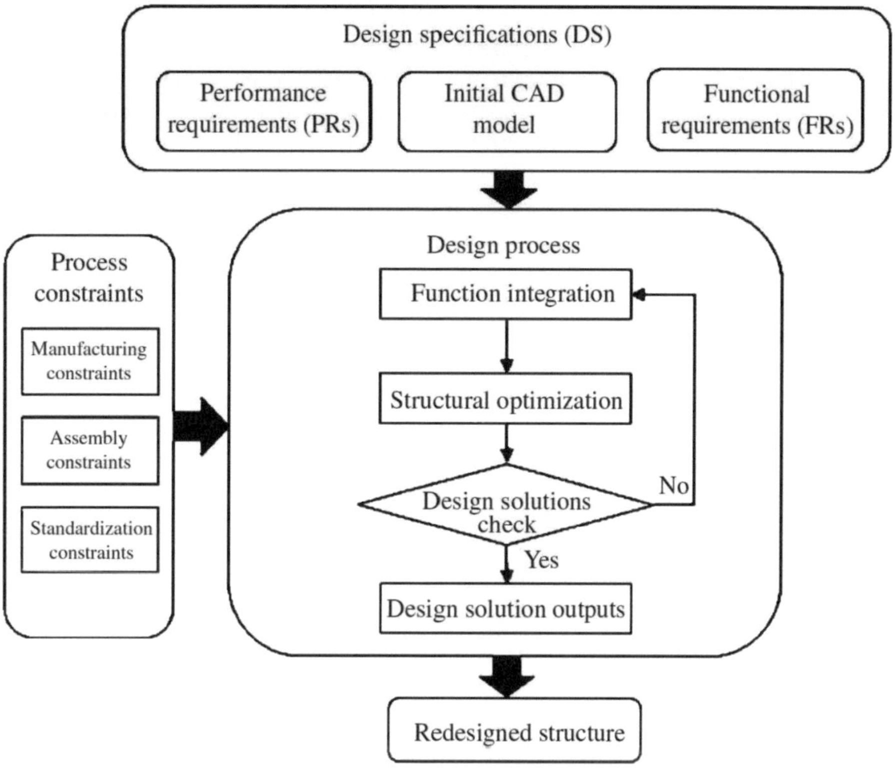

Figure 1.
Additive manufacturing-enabled design methodology. Source: Redrawn and adapted from [9].

Several researchers have applied DfAM in the design/redesign of automotive components. Bikas, Dalpadulo et al., Vaverka et al. [13–17] redesigned the wheel knuckles of formula race cars through topology optimization and considered laser powder bed fusion (LPBF). The redesign of a gearbox housing and automotive fixtures is seen in Barreiro et al. [18] and Naik et al. [19] respectively. For works focused on engines, Marchesi et al. [20] redesigned a diesel engine support while Barbieri et al. [21, 22] proposed that steel pistons could be used in place of their aluminum counterparts if they are "lightweighted" through topology optimization. In this chapter, topology optimization and strut- and sheet-based lattices are applied to redesign the connecting rod while making comparisons to a traditional design.

To combat expensive print runs to investigate the manufacturability of a part within the DfAM framework, developing a process model is important. A process model essentially predicts the response of a participating component or part by defining and solving the enabling physical phenomena usually through numerical or analytical mathematical approaches. In thermal-based powder-bed metal AM processes such as LPBF, being the most popular, complex multiphysics scenarios are involved. Some of the governing physics are mass transport within the meltpool, heat transfer by conduction, convection, and radiation, phase transformations, chemical reactions, Marangoni, capillary effects, etc. [7]. It is cumbersome and oftentimes impossible to consider all these physics in a high-fidelity process model to predict print success. Two major challenges in developing a high-fidelity process model are the ease of generating ideal, reliable, and validated mathematical solutions that incorporate all physics and the computational cost involved in running the models. While there have been decent attempts at developing full-scale thermal, thermo-mechanical, and thermo-metallurgical models at micro, meso, and macro scales [23–25], the need for fast-predictive models is pertinent. A widely accepted and applied fast-predictive model to calculate part deflection in LPBF is the inherent strain model (ISM) first developed by Ueda [26]. It requires strain values (often called inherent strains) calibrated from experiments run with exact process parameters intended for the actual part. These calibrated inherent strain values are utilized in a layer-by-layer pure elastic mechanical simulation to predict in-situ deformation and residual stresses within the part [27–29].

In **Figure 2**, a schematic diagram shows the typical workflow for ISM in a layerwise AM process. Initially, the base plate or substrate has zero displacements, strains, and stresses. When the first layer is deposited by the scan of a laser or electron beam on the powder bed, in the case of powder-bed processes, the experimentally calibrated inherent strain values are recorded to represent the load effects offered by the initial scanning process. With these strain values, equivalent force, displacement,

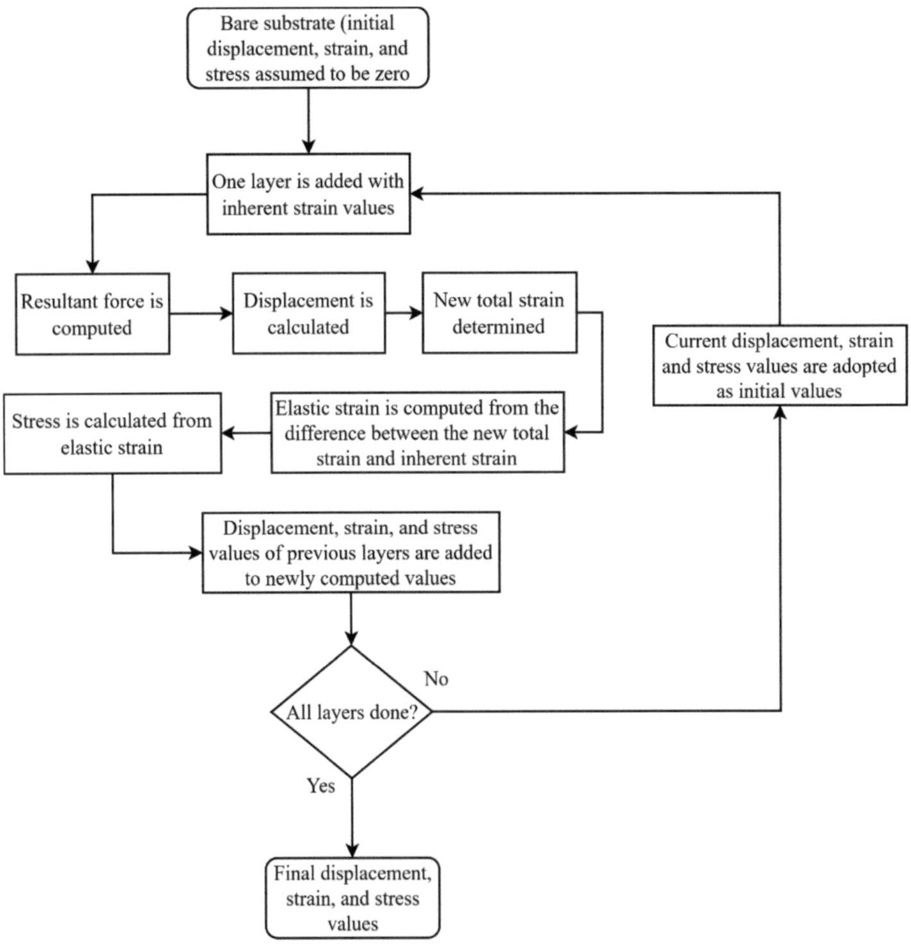

Figure 2.
Typical workflow of the inherent strain model for the prediction of deformation and residual stress in thermal-based metal AM processes.

strain, and stress vectors can be calculated. The calculated strain is referred to as the total strain with contributions from the elastic, plastic, thermal, and phase transformation strains. The inherent strain is a total of all strains except the elastic strain. Since ISM is strictly an elastic mechanical analysis, the elastic strain is obtained by subtracting the inherent strain from the total strain. The calculated stress, equivalent to the residual stress, is based on this elastic strain. The newly computed responses (displacement, strain, and stress) are added to the initial responses. Just before the analysis for the second layer commences, the summed responses are adopted as the initial responses and the whole workflow is carried out in this iterative manner till the last layer is done. To evaluate manufacturability, the maximum absolute deformation and/or residual stress are compared with acceptable values (e.g., yield/ultimate strength of the material for residual stress). For unacceptably high values, a redesign of the part can be done to either mitigate the deformation or residual stress or both. In this study, process simulations using ISM are carried out on all the connecting rod structures investigated and their maximum deformations are compared before and after support structure removal.

3. Topology optimization and lattice design

Topology optimization is a mathematical approach that finds the best structural layout of a design problem based on an objective and subject to one or more constraints [30–32]. It is one of three major structural optimization methods with size and shape optimization being the others. It is the preferred structural optimization method because it relies the least on the user's or designer's experience. While an initial structural configuration must be done for size and shape optimization, only the external structural boundary is required for topology optimization. More recently, topology optimization has given rise to generative design which performs the optimization without a strict single initial boundary or objective function. Nonetheless, defining the loads, boundary conditions, optimization objective(s), constraints, and material model is common to all optimization methods. There are several topology optimization methods: density-based, boundary/phase-field methods, evolutionary, etc. Several of them are based on the finite element method and a popular formulation to optimize structural rigidity is the compliance or strain energy objective formulation subject to a material volume constraint. Although the detailed formulation is not given here, a workflow of the optimization process is shown in **Figure 3**.

Another important tool in the DfAM framework is lattice design. A lattice structure can briefly be defined as one that comprises interconnected struts, or unit sheet features that are uniformly, pseudo-uniformly, or randomly patterned resulting in a cellular-like or meta-like design. Typically, for a lattice structure to be formed, the unit cell type, cell parameters (size, thickness, etc.), and lattice framework should be known. There are a host of benefits when lattice structures are used. They possess high strength-to-weight ratios, good heat transfer capabilities (especially surface-based lattices), negative Poisson's ratio, enhanced osseointegration capabilities (the functional and structural connection between a bone and an implant) when used for bio-implants, and low thermal expansivities. For their application in the re-design of the connecting rod in an ICE, their high strength-to-weight ratio is leveraged.

Three design iterations using the two tools outlined in previous paragraphs are performed. The original and initial designs are shown in **Figure 4a** and **b** respectively. **Figure 4c** and **d** show the two load cases considered for the structural design where

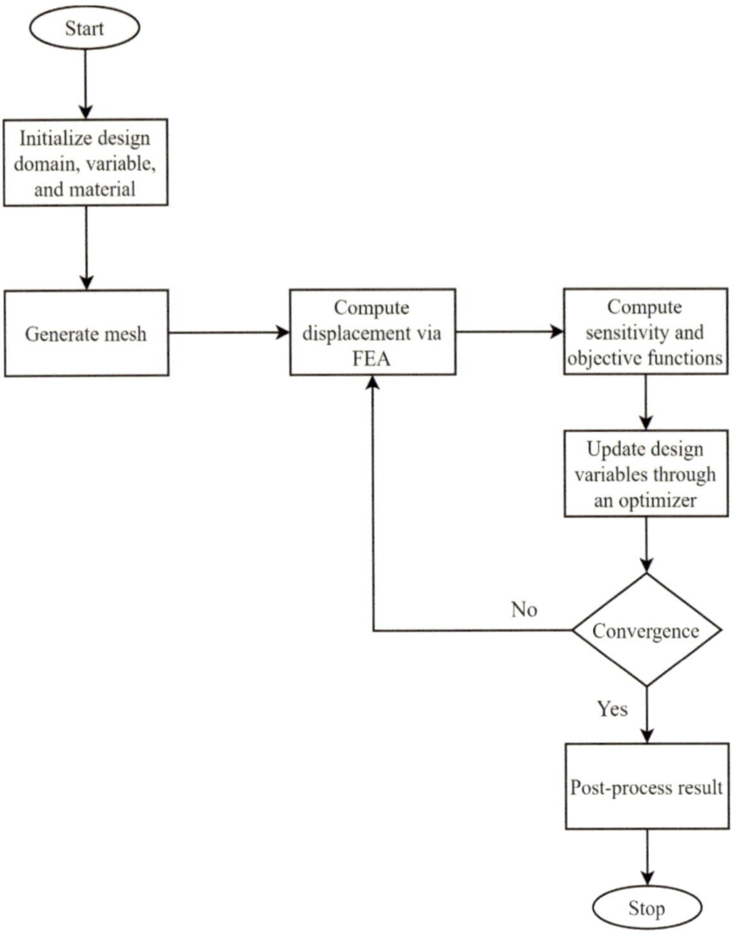

Figure 3.
A schematic showing the main sections of the topology optimization process.

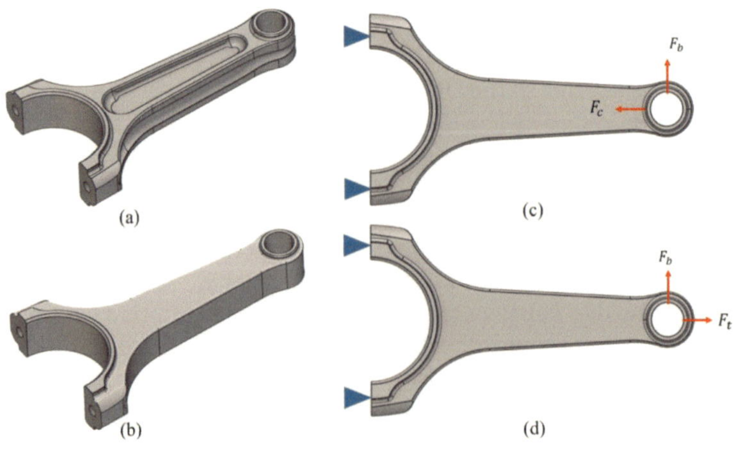

Figure 4.
Connecting rod showing the (a) original design, (b) initial design, (c) load case 1, and (d) load case 2.

F_c, F_t, and F_b are the compressive, tensile, and bending forces respectively. **Table 1** shows the design material specifications and load values as obtained by Alkalla et al. [33]. The original design was obtained from Grabcad [34] while the initial design is a simpler model that excludes some fine details in the original design. Typically, when performing topology optimization, a larger and simpler initial design domain is recommended.

All design workflows are performed in nTopology (academic license) [35]. There are some open source codes [36–38] with varying capabilities of performing topology optimization, however, nTopology offers the development of lattice structures with unique interfacing capacities with topology optimization.

3.1 Topology optimization (TO) design strategy

The first design workflow is a strict topology-optimized connecting rod, shown in **Figure 5**. Like most structural design workflows, the initial design domain is specified along with loads, fixed locations, and preserved regions. A finite element analysis (FEA) is then performed (static analysis in this case) to enable the computation of the objective and constraint(s). The objective is compliance minimization (tantamount to stiffness maximization) while a material volume constraint is imposed at a minimum of 40% and a maximum of 60%. All the information collected is then fed into the optimization module which generates an optimized model after a set number of iterations. This optimized model is usually made up of jagged features, therefore, a post-processing smooth step is established to finalize the model.

3.2 Topology optimization and lattice (TO-L) design strategy

The second design workflow is a combination of topology optimization and lattice design using a graded lattice approach. In **Figure 6**, there are two color schemes in the design workflow: light brown and blue. The light brown steps indicate topology optimization while the other indicates lattice design. It is observed that the topology optimization steps are the same as the previous design workflow except that the smoothened solution is utilized to generate an implicit thickness parameter field in this workflow. This thickness parameter is needed to establish the lattice gradation. While the thickness parameter is defined, a region that determines where the lattice

Parameter	Value
Material	Aluminum 7075-T6
Density	2700 kg/m^3
Elastic modulus	70 GPa
Yield strength	450 MPa
Poisson ratio	0.33
F_c, F_t	4246 N
F_b	1061 N

Table 1.
Material specification and load values.

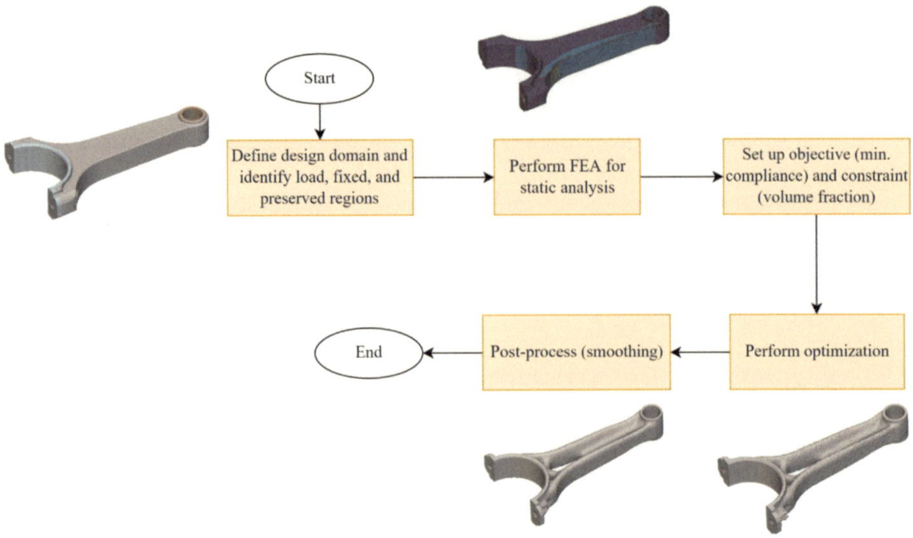

Figure 5.
Design workflow of a connecting rod through topology optimization. TO—topology optimization.

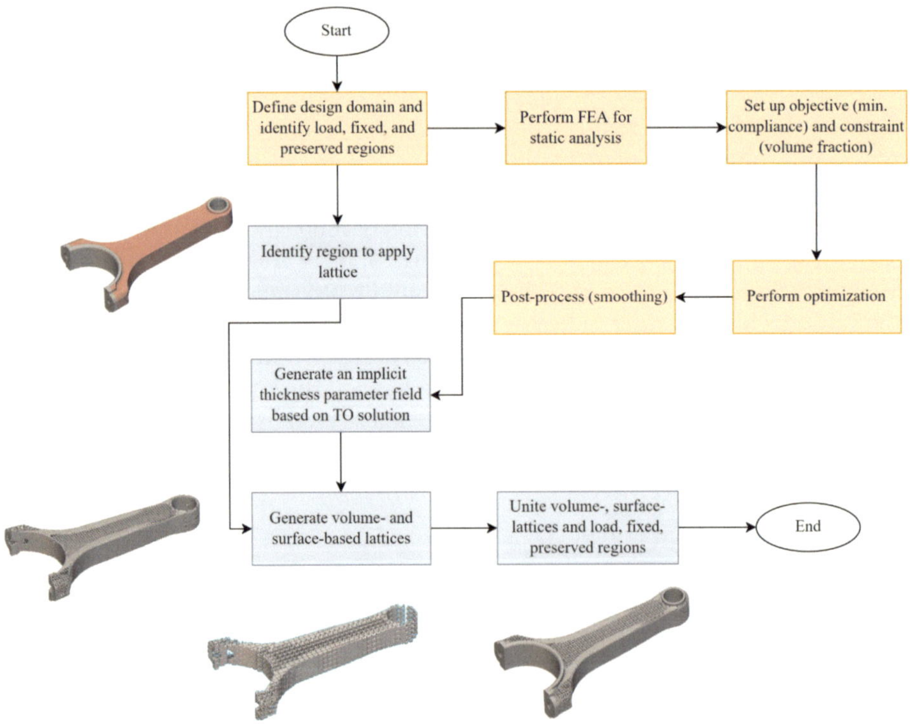

Figure 6.
Design workflow of a connecting rod through topology optimization and lattice design (graded lattice strategy). TO—topology optimization.

will be populated must be generated. The thickness parameter and the defined lattice region are both utilized to generate surface and volume strut lattices which are then combined to form the final graded lattice.

3.3 Topology optimization and lattice (TO-L) shell-infill design strategy

The third workflow also utilizes topology optimization and lattice design but in a shell-infill approach as displayed in the workflow in **Figure 7**. The smoothened model obtained from topology optimization undergoes two steps: an outer shell is created and the difference between the outer shell and smoothened model (a core) is obtained creating a shell-core model. A lattice operation is done on the solid core model for transformation to a strut- or surface-based lattice. In this work, two lattice cells are used: a strut-type body-centered cubic (BCC) and a gyroid. The shell and lattice structures are unioned to obtain the final shell-infill connecting rod as shown in **Figure 7**. The final shell-infill models of the connecting rod in **Figure 7** show only half of the shells to reveal the infill structures.

4. AM process simulation

As highlighted in Section 2, ISM is used to predict the deformation and stress response of the connecting rod designs during the build process. LPBF is adopted as the AM process in this study and AlSi10Mg is used as material. For process parameters, a power of 200 W, a scan speed of 1 m/s scan, and a layer thickness of 30 μm were used. Based on these parameters, two cantilevers were printed and the vertical deformations of the cantilever tips after cutting were recorded as 2 mm each. The printing process and details of calibrating the inherent strain values are beyond the scope of this chapter, and similar studies can be seen in [39, 40]. The inherent strain values used in this study were generated by the Simufact Additive®'s metal AM

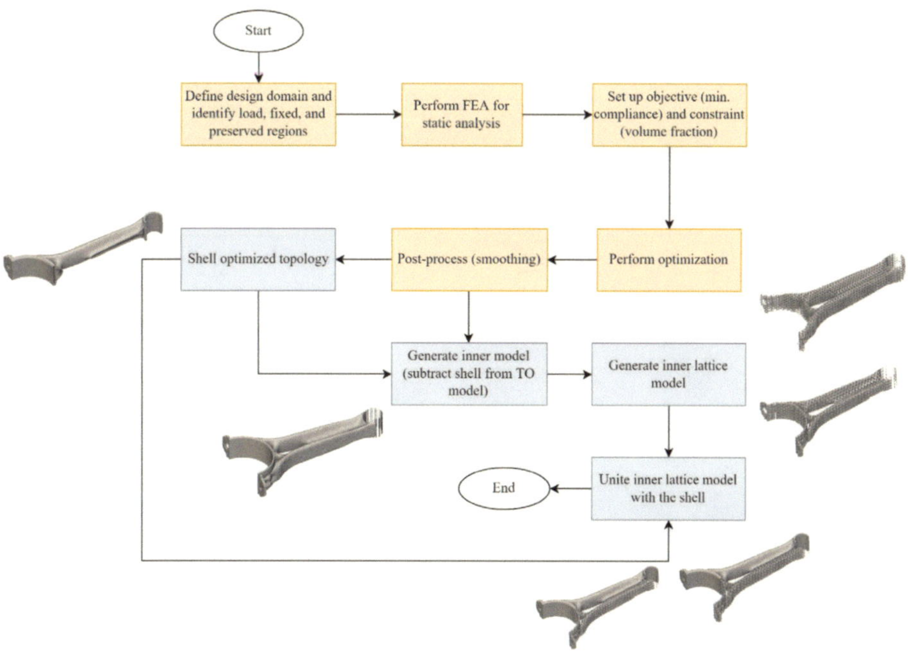

Figure 7.
Design workflow of a connecting rod through topology optimization and lattice design (shell-infill strategy). TO—topology optimization.

calibration study and were found to be $\varepsilon_x = -0.00292, \varepsilon_y = -0.00297, \varepsilon_z = -0.03$ based on the deformation values from the cantilever experiments. Simufact Additive® was also used to solve the elastic mechanical problem for all designs based on ISM.

5. Original vs optimized designs

5.1 Structural performance and material savings

Static analysis was done on all optimized models and a benchmark design. The original and optimized models are shown in **Figure 8**. Static analysis was performed on all designs according to **Figure 4c** and **d** and the von Mises (VM) equivalent stress distributions are shown in **Figure 9**. For all the designs, the maximum VM stresses are toward the crank pin end, and they all possess similar stress distributions. For the optimized designs, the graded lattice strategy is the most complicated to implement in nTopology while the topology optimized model was the fastest to realize. The reason for this is the dependence of the lattice strategies on the post-processed topologically optimized model.

To analyze the optimality of the structures for material and function, the volume fractions, maximum displacements, maximum VM stresses, and factor of safeties are investigated and compared. The volume fraction is the ratio of the material volume of a new design to the original/benchmark design. The maximum displacement and VM

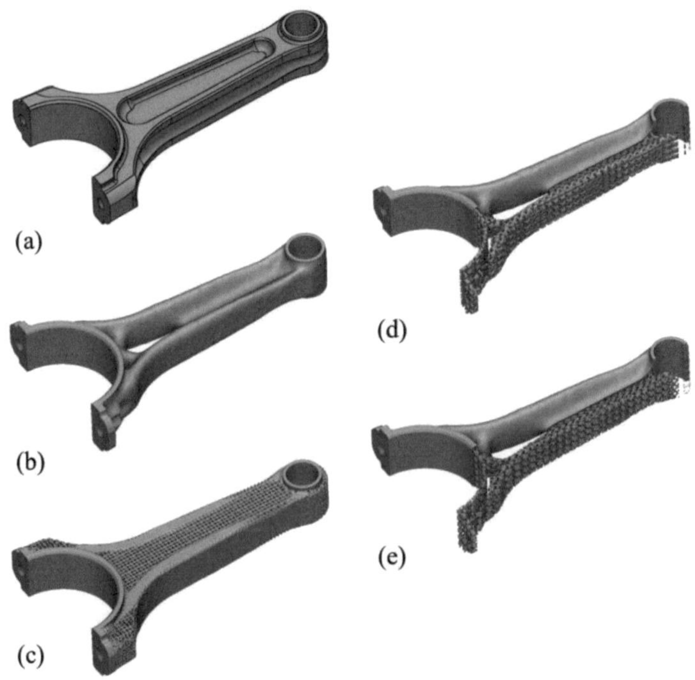

Figure 8.
Versions of the connecting rod (a) original (b) topology optimized (c) topology optimized and latticed (graded) (d) topology optimized shell-lattice infill (BCC) (e) topology optimized shell-lattice infill (gyroid) design NB: the shell of the designs in (d) and (e) is sectioned along the longest axis to reveal the lattice infill.

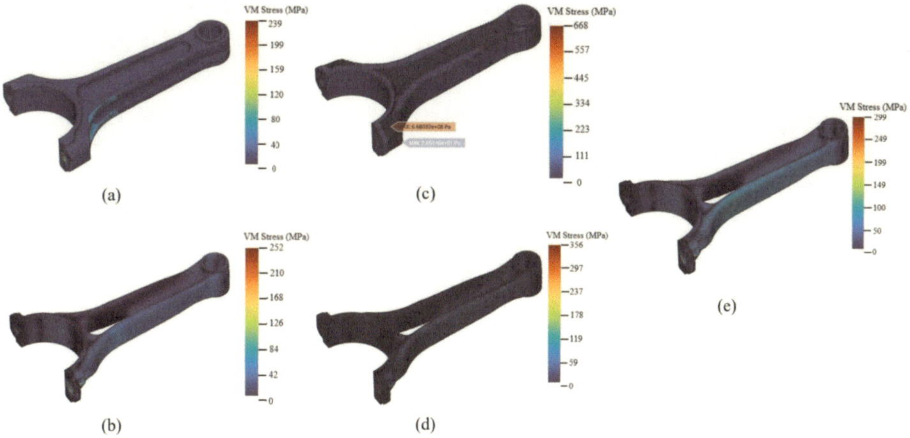

Figure 9.
VM stress of the (a) original (b) topology optimized (c) topology optimized and latticed (graded) (d) topology optimized shell-lattice infill (BCC) (e) topology optimized shell-lattice infill (gyroid) design.

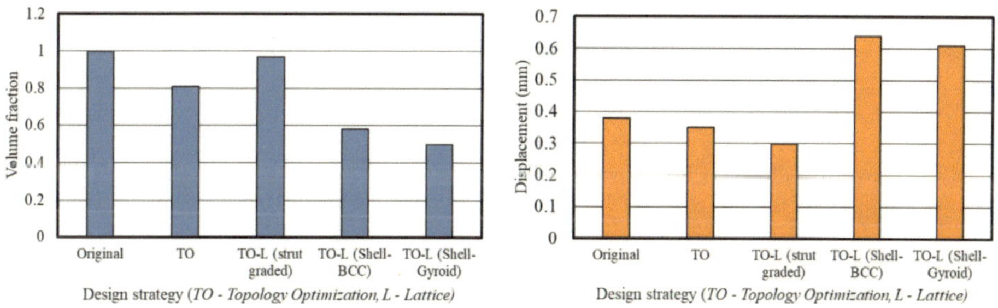

Figure 10.
Final volume fraction of design (left image), maximum displacement from static analysis (right image).

stresses are the maximum nodal values of these quantities from static FEA while the factor of safety for ductile materials is utilized. **Figures 10** and **11** show how the volume fraction, maximum displacement, maximum VM stress, and factor of safety vary from design to design. Typically, the lower the volume fraction, the better the design as less material is used. The higher the VM stress and displacement, the poorer the design while higher factors of safety are favorable.

The topology-optimized-lattice (TO-L) graded design is almost equal in material amount as the original/benchmark design while the TO-L shell-gyroid design uses the least material volume. However, the shell-infill designs give considerably higher maximum displacements relative to the others, about twice higher. It is interesting to know that although the TO and TO-graded lattice designs utilize less material volume, they give lower maximum displacements compared to the original design. In **Figure 11**, the maximum VM stresses and factor of safety are given. The TO-L graded design has almost twice the maximum VM stresses of that of the second-placed design (shell-BCC). Moreover, it gives a factor of safety below 1 which signifies the structure is inadequate even from a static design viewpoint. The graded lattice design can be improved by investigating other cell structures and lattice thickness fields. Unfortunately, since the graded design approach also takes more man-hours to implement, it is less attractive compared to the other approaches.

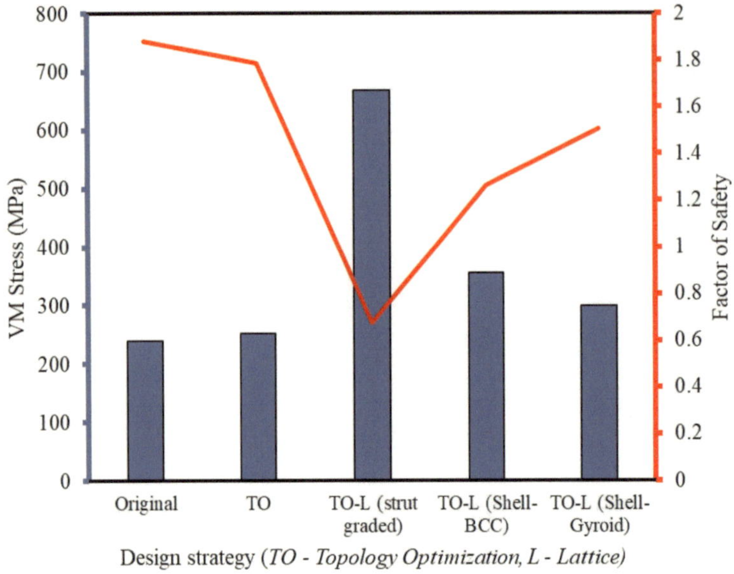

Figure 11.
Maximum von Mises stress and factor of safety of all designs.

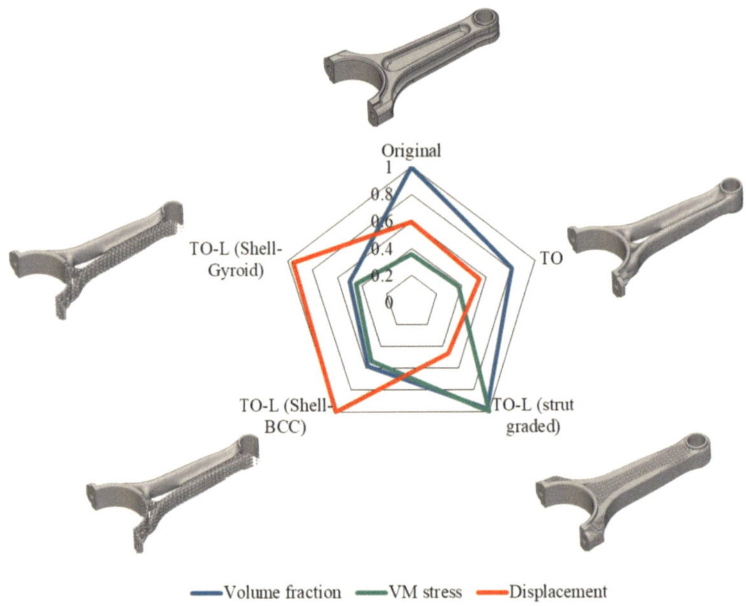

Figure 12.
Radar plot of all designs comparing volume fractions, normalized VM stresses, and normalized maximum displacements.

The final image in **Figure 12** shows a radar plot that quite nicely compares the three major design criteria: volume fraction, normalized VM stresses, and normalized maximum displacements. The figure immediately shows that the TO model is comparable in performance to the original design although with almost a 20% reduction (according to **Figure 10**) in material usage. However, if there are no strict requirements on the maximum VM stress and displacement, the shell-infill strategies are the best approaches since they result in models that use much less material (about 50% less than the original) while keeping their factor of safety above 1.

5.2 Manufacturability

In this study, residual stress and deformation before and after support removal form the basis of assessing the manufacturability of all designs. The maximum residual stresses for all designs were 370 MPa, and the plots are not shown because of this homogeneity, also, no further assessments are made to judge comparative manufacturability based on the stresses. The displacement plots after support removal are shown in **Figure 13** and all designs show quite similar distributions except the topology-optimized and latticed (TO-L) design. The TO-L design shows higher

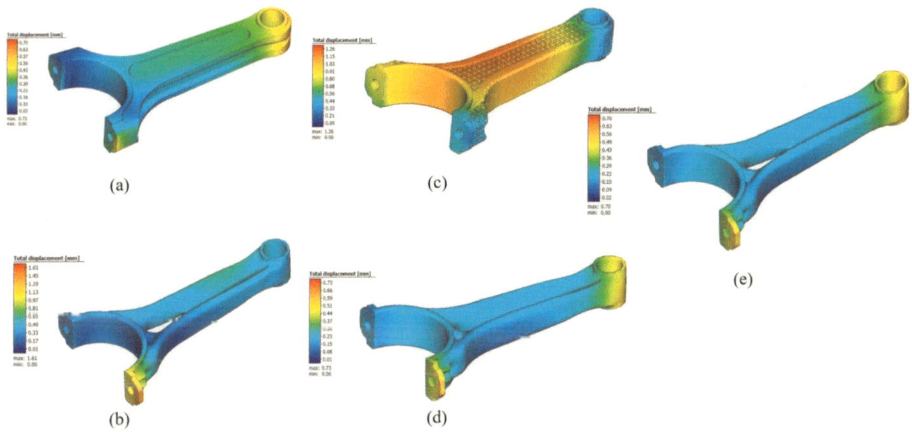

Figure 13.
AM process-induced displacements after support removal for the (a) original (b) topology optimized (c) topology optimized and latticed (graded) (d) topology optimized shell-lattice infill (BCC) (e) topology optimized shell-lattice infill (gyroid) design.

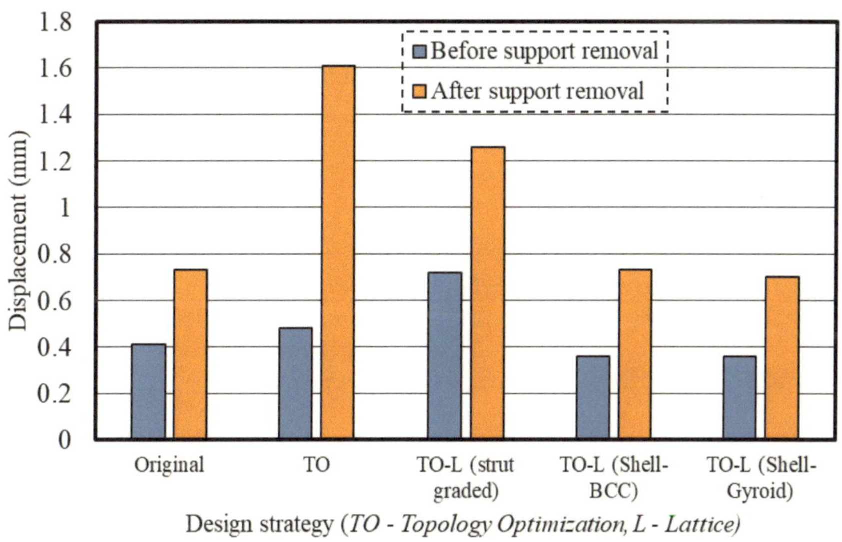

Figure 14.
Maximum AM process-induced displacements in the designs before and after support removal.

displacements around the middle portion of the connecting rod as opposed to the other designs where they appear at both ends. **Figure 14** reveals that the TO design offers the maximum deformation of all designs at 1.6 mm after support removal. The TO-L design also gives a maximum deformation of over 1.2 mm after support removal while the other designs are at around 0.7 mm. It is interesting to note that the TO-L shell-infill strategies give deformations similar to the original design before and after support removal while noting that they both utilize approximately 50% less material. It should be noted that these results are peculiar to the process parameters and build orientation. The deformations can potentially reduce when these parameters are optimized. Moreover, if build orientation optimization is performed for each design type, different build angles might be realized possibly influencing the final deformations.

6. Conclusion

To achieve lower carbon footprints in ICEs, reducing the weight of the components is paramount. Additive manufacturing, new design tools, and techniques can help to attain this. In this study, the use of topology optimization and lattice structures to redesign a connecting rod has shown success in this direction. Also, with the advancement of structural design tools such as nTopology, which uses field- and implicit-driven modeling concepts, these new design pathways can be implemented quite easily. This study shows that the integration of topology optimization and lattice design through a shell-infill approach can be adopted to redesign a connecting rod with up to 50% material savings while keeping the structural performance in static conditions at an acceptable level. Also, results from the ISM-based AM process simulation show lower deformations in the shell-infill designs before and after support removal compared to other designs. It should be noted, however, that the overall performance of a connecting rod, like several other components in an ICE, is dictated by other parameters/conditions such as fatigue life, damage, eigenfrequencies, etc. It is therefore recommended that further analysis be carried out on the model outcomes of these nascent design approaches for either validation or model calibration. Moreover, a broad design of experiment study related to build orientation optimization for the AM process should be investigated. Moreover, lattice-alone designs and varying lattice cell parameters should be explored in the future.

Author details

Osezua Ibhadode
Multifunctional Design and Additive Manufacturing (MDAM) Lab, Department of Mechanical Engineering, University of Alberta, Edmonton, Alberta, Canada

*Address all correspondence to: ibhadode@ualberta.ca

References

[1] Paolo Carlucci A. Introductory chapter: The challenges of future internal combustion engines. Future International Combustion Engines. 2019;**2019**:1-4

[2] Czerwinski F, Amirkhiz BS. On the Al – Al 11 Ce 3 eutectic transformation in Ce 3 eutectic transformation in alloys. Materials (Basel). 2020;**13**:4549

[3] Czerwinski F. Thermal stability of aluminum-nickel binary alloys containing the Al-Al3Ni eutectic. Metallurgical and Materials Transactions A, Physical Metallurgy and Materials Science. 2021; **52**(10):4342-4356

[4] Czerwinski F. Current trends in automotive lightweighting strategies and materials. Materials (Basel). 2021;**14**(21): 1-27

[5] Ibhadode A, Ebhojiaye R. A new lightweight material for possible engine parts manufacture. In: The Future of Internal Combustion Engines. London, UK: Intechopen; 2018. p. 19

[6] Rosen DW. Computer-aided design for additive manufacturing of cellular structures. Computer Aided Design Applications. 2007;**4**(5):585-594

[7] Toyserkani E, Sarker D, Ibhadode OO, Liravi F, Russo P, Taherkhani K. Metal Additive Manufacturing. Hoboken, New Jersey: Wiley; 2022

[8] Tang Y, Hascoet J-Y, Zhao YF. Integration of topological and functional optimization in design for additive manufacturing. In: Proc. ASME 2014 12th Bienn. Conf. Eng. Syst. Des. Anal. Copenhagen, Denmark. 2015. pp. 1-8

[9] Yang S, Zhao YF. Additive manufacturing-enabled design theory and methodology: A critical review. International Journal of Advanced Manufacturing Technology. 2015; **80**(1-4):327-342

[10] Primo T, Calabrese M, Del Prete A, Anglani A. Additive manufacturing integration with topology optimization methodology for innovative product design. International Journal of Advanced Manufacturing Technology. 2017;**93**(1-4):467-479

[11] Orquéra M, Campocasso S, Millet D. Design for additive manufacturing method for a mechanical system downsizing. Procedia CIRP. 2017;**60**: 223-228

[12] Lei N, Yao X, Moon SK, Bi G. An additive manufacturing process model for product family design. Journal of Engineering Design. 2016;**27**(11):751-767

[13] Bikas H, Stavridis J, Stavropoulos P, Chryssolouris G. A design framework to replace conventional manufacturing processes with additive manufacturing for structural components: A formula student case study. Procedia CIRP. 2016; **57**:710-715

[14] Dalpadulo E, Pini F, Leali F. Integrated CAD platform approach for design for additive manufacturing of high performance automotive components. International Journal on Interactive Design and Manufacturing. 2020;**2020**:899-909

[15] Walton D, Moztarzadeh H. Design and development of an additive manufactured component by topology optimisation. Procedia CIRP. 2017;**60**: 205-210

[16] Reddy SN, Maranan V, Simpson TW, Palmer T, Dickman CJ.

Application of topology optimization and design for additive manufacturing guidelines on an automotive component. Proceedings of the ASME Design Engineering Technical Conference. 2016;**2A-2016**:1-10

[17] Vaverka O, Koutny D, Palousek D. Topologically optimized axle carrier for formula student produced by selective laser melting. Rapid Prototyping Journal. 2019;**25**(9):1545-1551

[18] Barreiro P, Bronner A, Hoffmeister J, Hermes J. New improvement opportunities through applying topology optimization combined with 3D printing to the construction of gearbox housings. Forsch Ingenieurwes. 2019;**83**:669-681. DOI: 10.1007/s10010-019-00374-1

[19] Naik A, Sujan T, Desai S, Shanmugam S. Light-weighting of additive manufactured automotive fixtures through topology optimization techniques, SAE Tech. Pap., no. 2019

[20] Marchesi TR et al. Topologically optimized diesel engine support manufactured with additive manufacturing. IFAC-PapersOnLine. 2015;**48**(3):2333-2338

[21] Barbieri SG, Giacopini M, Mangeruga V, Mantovani S. A design strategy based on topology optimization techniques for an additive manufactured high performance engine piston. Procedia Manufacturing. 2017;**11**(June): 641-649

[22] Barbieri SG, Giacopini M, Mangeruga V, Emilia R. Design of an additive manufactured steel piston for a high performance engine: Developing of a numerical methodology based on topology optimization techniques. 2018; pp. 1-10

[23] Kouraytem N, Li X, Tan W, Kappes B, Spear AD. Modeling process–structure–property relationships in metal additive manufacturing: A review on physics-driven versus data-driven approaches. Journal of Physical Materials. 2021;**4**(3):32002

[24] Bayat M, Dong W, Thorborg J, A. C. To, Hattel JH. A review of multi-scale and multi-physics simulations of metal additive manufacturing processes with focus on modeling strategies. Additive Manufacturing. 2021;**47**:102278

[25] Hashemi SM et al. Computational modelling of process–structure–property–performance relationships in metal additive manufacturing: A review. International Materials Review. 2021:1-46

[26] Ueda Y, Kim YC, Yuan MG. A predicting method of welding residual stress using source of residual stress. Journal of Japan Welding Society. 1988; **6**(1):59-64

[27] Keller N, Ploshikhin V. New method for fast predictions of residual stress and distortion of AM parts. In: 25th Annual International Solid Freeform Fabrication Symposium &#65533; An Additive Manufacturing Conference, SFF 2014. Vol. 2014. pp. 1229-1237

[28] Siewert M, Neugebauer F, Epp J, Ploshikhin V. Validation of mechanical layer equivalent method for simulation of residual stresses in additive manufactured components. Computer Mathematics with Applicatios. 2019; **78**(7):2407-2416

[29] A. Yaghi, S. Ayvar-Soberanis, S. Moturu, R. Bilkhu, and S. Afazov, "Design against distortion for additive manufacturing," Conference Additives Manufacturing Benchmarks, vol. 27, no. March, pp. 224-235, May 2018

[30] Bendsøe MP, Sigmund O. Topology Optimization: Theory, Methods, and

Applications. 2nd ed. Berlin, Heidelberg. 2003

[31] Ibhadode O, Zhang Z, Rahnama P, Bonakdar A, Toyserkani E. Topology optimization of structures under design-dependent pressure loads by a boundary identification-load evolution (BILE) model. Structural and Multidisciplinary Optimization. 2020;**62**(4):1865-1883

[32] Zhang ZD et al. Topology optimization parallel-computing framework based on the inherent strain method for support structure design in laser powder-bed fusion additive manufacturing. International Journal of Mechanics and Materials in Design. 2020;**2020**:0123456789

[33] Alkalla MG, Helal M, Fouly A. Revolutionary superposition layout method for topology optimization of nonconcurrent multiload models: Connecting-rod case study. International Journal for Numerical Methods in Engineering. 2021;**122**(5):1378-1400

[34] Yüksel A. Titanium Connecting Rod and Piston. Grabcad. 2022. [Online]. Available: https://grabcad.com/library/titanium-connecting-rod-and-piston-1

[35] nTopology. "nTopology." 2022

[36] Ibhadode O, Zhang Z, Bonakdar A, Toyserkani E. IbIPP for topology optimization - an image-based initialization and post-processing code written in MATLAB. SoftwareX. 2021; **14**:100701

[37] Zhang ZD, Ibhadode O, Bonakdar A, Toyserkani E. TopADD: A 2D/3D integrated topology optimization parallel-computing framework for arbitrary design domains. Structural and Multidisciplinary Optimization. 2021; **2021**:1701-1723

[38] Aage N, Andreassen E, Lazarov BS. Topology optimization using PETSc: An easy-to-use, fully parallel, open source topology optimization framework. Structural and Multidisciplinary Optimization. 2015; **51**(3):565-572

[39] Hovig EW, Azar AS, Mhamdi M, Sørby K. Mechanical properties of AlSi10Mg processed by laser powder bed fusion at elevated temperature. In: BT - TMS 2020 149th Annual Meeting & Exhibition Supplemental Proceedings. San Diego, California. 2020. pp. 395-404

[40] Cheng L, Liang X, Bai J, Chen Q, Lemon J, A. To. On utilizing topology optimization to design support structure to prevent residual stress induced build failure in laser powder bed metal additive manufacturing. Additive Manufacturing. 2019;**27**(March): 290-304

Chapter 3

Improved Smell Agent Optimization Sizing Technique Algorithm for a Grid-Independent Hybrid Renewable Energy System

Akawu Shekari Biliyok and Salawudeen Ahmed Tijani

Abstract

This chapter discuss an improvement on the novel computational intelligent algorithm using the smell phenomenon. In the standard smell agent optimization algorithm, the olfactory capacity is constant thereby assuming that every smell agent has the same sensing capacity. In the improved smell agent optimization algorithm, that is changed to account for the difference in smell agent capacity. The algorithm was run against the standard smell agent optimization on Matlab to find the best HRES design using annual cost, Levelized cost of electricity (LCE), loss of power supply probability (LPSP) and excess energy. It was shown after the comparative analysis that there was a 79%, 99.9% and 53.4% improvement for annual cost, LCE and LPSP respectively. Statistically, results showed that the iSAO obtained the most cost effective HRES design compared to the benchmarked algorithms.

Keywords: smell agent optimization (SAO), improved smell agent optimization (iSAO), hybrid renewable energy system (HRES), levelized cost of electricity (LCE), loss of power supply probability (LPSP)

1. Introduction

Technological and societal developments can be observed as the drivers of the changes in interconnected systems like Mini-grids or hybrid renewable energy systems (HRES). HRES consists of two or more renewable sources used to provide system efficiency as well as create better balance in energy tied to the conventional grid or off-grid with battery storage [1, 2] and there are various optimization sizing techniques for such systems namely;

i. Dynamic programming.

ii. Graphical construction technique.

iii. Probabilistic approach.

iv. Multi-objective design.

v. Linear programming.

vi. Iterative technique.

vii. Artificial intelligence.

Over the years, optimization has developed into an established field of computation intelligence (CI) inspired by various natural behavioral rules. Some of these natural behavioral based optimization methods of imitating evolution, ecology, animal activities and apparatus of human culture were established to ease solving several categories of complicated social, economic, scientific and engineering design problems. Such as:

i. Weak problems with little or no area information.

ii. Problems for which a near-ideal solution can be satisfactory.

iii. Non-deterministic Polynomial (NP) complete problems.

iv. Problems with non-smooth and noisy search space.

v. Problems whose environments are uncertain/changes or both [3, 4].

Each branch has a comprehensive theoretical basis and is highlighted by a collection of sophisticated algorithms and software like Genetic Algorithm (GA) which is a heuristic search technique used in artificial intelligence and computing. In this approach, evolution is performed through Elitism, Crossover, Mutation [5].

Artificial intelligence (AI) techniques are suitable as substitute methods to conventional techniques or as mechanisms of integrated systems. These are applied in solving complicated practical problems in various areas and are popular nowadays. AI-techniques have the following features;

i. They learn from previous examples.

ii. They are fault-tolerant and can handle noisy and incomplete data.

iii. They deal with non-linear problems.

iv. Once trained, they can perform forecasts at high speed.

The smell agent optimization technique (SAO) is a newly developed meta-heuristic algorithm using the phenomenon of smell perceptions. The concept of SAO is developed in three distinct modes; sniffing, trailing and random mode [6].

One of the five senses through which the world is perceived is the sense of smell (olfactory). Through the sense of smell, humans and other animals can perceive a large number of chemicals in the external world which enables us to perceive the molecular concentration or smell and intuitively trace this concentration in order the identify the source [7]. In the conventional SAO, the olfaction capacity is set to a constant value which assumes that the smell agent sense of smell cannot change and this assumption comes with certain drawbacks like longer computational time.

2. Case study

This case study proposed an improved SAO technique algorithm that makes the olfaction capacity a dynamic variable thereby improving speed and precision of solving the combinatorial problem of selecting the optimal combination between a hybrid renewable energy system at best annual cost, loss of power supply probability (LPSP) and levelized cost of energy (LCE) (**Figures 1**, **5–10**).

A mathematical model may fail due to any or all of the following reasons:

i. The problem or process may be too complex for mathematical reasoning.

ii. The problem or process may be dynamic and stochastic in nature.

iii. The solution space of the problem may be too large for mathematical computation.

iv. The problem or process may contain some uncertainties.

All these characteristics are exhibited by most real-life problems [8]. To address this challenge, researchers have employed computational intelligence techniques under different environments and promising results have been achieved.

To achieve the stated aim, the following objectives were adopted.

i. Collection of load data and weather data such as solar insolation, solar temperature and wind speed from questionnaires and the NASA website.

ii. To model the hybrid energy system considering PV/Battery, Wind/Battery, PV/Wind/Battery configurations and load.

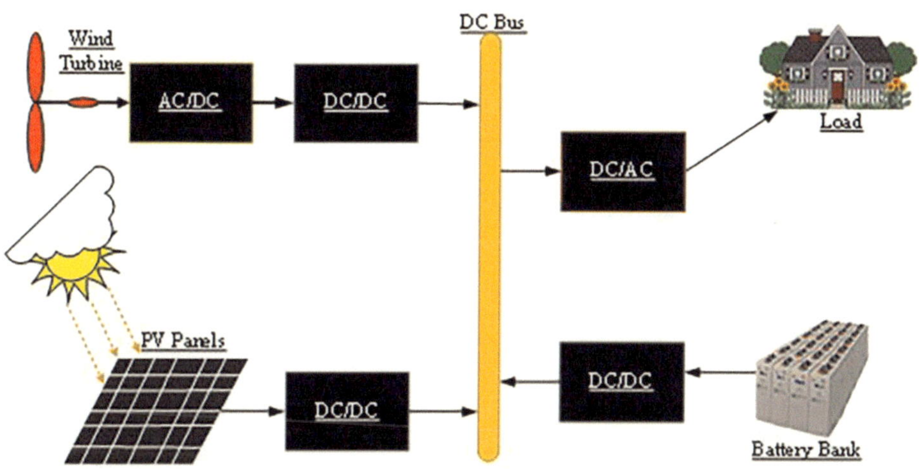

Figure 1.
Illustrates schematic diagram of the proposed hybrid renewable energy system [1].

iii. To develop an improved Smell Agent Optimization iSAO and optimize the developed hybrid model in 2, using the iSAO and the standard SAO.

iv. To validate by comparing the performance of iSAO with SAO on Matlab Simulink.

2.1 Significance of study

A lot of research has been done on efficient sizing utilizing artificial intelligence however, this research proposes an artificial intelligence smell agent optimizing sizing technique for a solar PV, wind turbine and battery storage system. The contributions are as follows;

i. This algorithm is based on the creation of a surface with smell trails and iteration of the agents in finding a path. It can be applied in various computational constraints that use path-based problems and this is useful in solving NP-hard constraints that are related to path discovery [9].

ii. The economic enticement of distribution companies around the world (DISCOs) is to minimize losses in the network. So, when real losses are greater than the standard losses, the DISCOs loses economically or profit when the opposite occurs. This loss minimization in distribution systems is well-suited for researchers [10].

iii. Improving the smell agent optimization sizing technique for a hybrid renewable energy system.

Load data was collected with the help of questionnaires to understand energy demand patterns and from a reputable independent power provider (IPP).

The algorithm was modeled for the following configurations:

i. PV, wind turbines and battery storage configurations.

ii. PV and battery storage configurations.

iii. Wind turbines and battery storage configurations.

In the end, the modified smell agent optimization technique was validated on Matlab Simulink and compared to the standard smell agent optimization technique with annual cost of maintenance, Levelized cost of energy, loss of power supply probability and excess energy as the unit of measure.

3. Smell agent optimization

The Smell Agent Optimization (SAO) is a recently developed algorithm from the idea of how a smelling agent learns to identify a smell source. The authors of the algorithm argue that, with a well-developed olfaction capacity, an organism (including humans) can perceive smell substance and intuitively follows the smell substance to identify its source [11, 12]. This idea is modeled into three classifiable modes

DOI: http://dx.doi.org/10.5772/intechopen.105489

namely; the sniffing, trailing and random modes. These modes interact together to form the smell agent optimization. The detailed information about these modes is discussed as follows:

i. Sniffing Mode

The idea of sniffing mode hinges on the ability of a smell agent (i.e. human) to first perceive the presence of smell molecules around its surrounding while the molecules constantly diffuse from its smell source. This analogy can be further portrayed as, when a smell molecule evaporates from a smell source in the direction of an agent, the agent sniff (sense or perceive) and decide whether it's a pleasant smell or a harmful smell. After this decision is made, the agent either moves towards the direction of the smell molecules or moves away from the smell molecules. Assuming N is the total number of smell molecules evaporating from a smell source and D is the size of the search spacing where the smell molecules have occupied (dimension). The initial positions of smell molecules can be generated as follows:

$$x_i^t = \left[x_{N,1}^t, x_{N,2}^t, \dots, x_{N,D}^t\right] \tag{1}$$

Meanwhile, the smell molecule's evaporation from the source is in Brownian form, to sustain this Brownian evaporation, each molecule is assigned an initial velocity to aid their movement in the search space. This is achieved using the following equation

$$v_i^t = \left[v_{N,1}^t, v_{N,2}^t, \dots, v_{N,D}^t\right] \tag{2}$$

From Eqs. (1) and (2), the evaporation of smell molecules can therefore be represented as follows:

$$x_i^{t+1} = x_i^t + \left(v_i^{t+1} \times \Delta t\right) \tag{3}$$

where v_i^{t+1} is the change in velocity and Δt is the change in time? Note that for the optimization process, change in time is always 1, i.e., an algorithm moves from one iteration to another with a constant step of 1.

For the molecules to evaporate from one point to another, the velocity of every molecule is updated as follows:

$$v_i^{t+1} = v_i^t + \alpha_1 \times \sqrt{\frac{3kT}{m}} \tag{4}$$

Thus,

$$x_i^{t+1} = x_i^t + v_i^{t+1} \tag{5}$$

where α_1 is a random number generator, k is Boltzmann's constant given by 1.38 x$10^{-23} JK^{-1}$, T and m are the temperature and mass of smell molecules respectively.

The agent snorts these smell molecules by estimating the fitness of Eq. (5) and decide whether to track the source of the smell molecules or not.

ii. Trailing Mode

From the implementation of the sniffing mode, the fitness of each smell molecule is evaluated, then the position of the agent x^t_{agent} and the position of worst molecules x^t_{worst} are determined as the positions of molecules with best and worst sniffing fitness respectively. These positions are used to model the trailing behavior of the agent towards the smell source as follows:

$$x_i^{t+1} = x_i^t + \alpha_2 \times \eta \times \left(x_i^t - x^t_{agent}\right) - \alpha_3 \times \eta \times \left(x_i^t - x^t_{worst}\right) \tag{6}$$

where η which is also called *olf* is the olfaction capacity of the agent, α_2 and α_3 are random numbers used to penalize the influence η x^t_{agent} on and x^t_{worst} respectively.

iii. Random Mode

The random mode, which is like a savior strategy the agent employed to avoid getting stuck in local minima. This mode only becomes active when the agent could not obtain the global solution to the optimization problem after implementing the trailing mode. The model is expressed as:

$$x_i^{t+1} = x_i^t + \alpha_4 \times \Phi \tag{7}$$

where Φ a constant step taking by the agent andα_4 is a random number used to penalize the influence of Φ.

3.1 Objective function formulation

The objective function considered in this work is a tri-objective optimization problem where the aim is to minimize the Levelized Cost of Energy (LCOE), Loss of Power Supply Probability (LPSP) and Excess Energy Generated. This is expressed in Eq. (8):

$$\text{minf} = \min\ (\curlywedge\text{LCOE} + \curlywedge\text{LPSP} + \curlywedge\text{EE}) \tag{8}$$

Where: $\curlywedge$ *is penalty factor*.
These individual objective functions can be expressed as follows:

$$\text{LCOE} = \frac{\text{C}_{\text{A\_total}}}{\text{E}_{\text{total}}} \tag{9}$$

$$\text{LPSP} = \frac{\sum_0^{\text{T}} \text{P}_{\text{Load}} - \text{P}_{\text{pv}} - \text{P}_{\text{wind}} - \text{P}_{\text{SOC}_\text{M}}}{\sum_0^{\text{T}} \text{P}_{\text{Load}}} \tag{10}$$

$$\text{EE} = \sum_0^{\text{T}} \frac{\text{P}_{\text{Load}} - \text{P}_{\text{pv}} - \text{P}_{\text{wind}}}{\text{P}_{\text{pv}} + \text{P}_{\text{wind}}} \tag{11}$$

where; C_{A_total} is the total cost of the hybrid systems, E_{total} is the total cost of energy generation, P_{Load} is the load demand, P_{pv} is the output power from the PV generation units, P_{wind} is the output power from the wind turbine generation unit and T is the total time.

To achieve these tri-objectives, the total annual cost given in equation Eq. (12) must also be minimized.

$$\mathrm{TAC} = \sum_{i=1}^{N} \mathrm{AMC} + \sum_{i=1}^{N} \mathrm{ACC} \tag{12}$$

where N is the total hours considered, TAC is the total annual cost, AMC is the annual maintenance cost and ACC is the annual capital cost. The annual maintenance cost is expressed as:

$$\mathrm{AMC} = \mathrm{n_{pv}P_{pvm}} + \mathrm{n_{wt}P_{wtm}} \tag{13}$$

Whereas, the total capital cost is calculated as:

$$\mathrm{ACC} = \mathrm{CFR} \times \left[\mathrm{n_{pv}C_{pv}} + \mathrm{n_{wt}C_{wt}} + \mathrm{n_{Bat}C_{Bat}} + \mathrm{n_{Inv}C_{Inv}}\right] \tag{14}$$

where n_{pv} is the number of PV panels, C_{pv} is the unit cost of PV panel, n_{wt} is the number of wind turbines, C_{wt} is the unit cost of a wind turbine, n_{Bat} is the number of batteries, C_{Bat} is the present worth of battery, n_{Inv} is the number of converters/inverters, C_{Inv} is the present worth of converter/inverter and CRF is the Capital Recovery Factor.

The system implemented for hybrid system design consisting of PV/Wind/Battery configuration. The order configurations considered in the study include PV/Battery and Wind/Battery. For PV/Battery systems design, the wind turbine generators shut down and equation Eqs. (15) and (16) is modified as:

$$\mathrm{AMC} = \mathrm{n_{pv}P_{pvm}} \tag{15}$$

Whereas, the total capital cost is calculated as:

$$\mathrm{ACC} = \mathrm{CFR} \times \left[\mathrm{n_{pv}C_{pv}} + \mathrm{n_{Bat}C_{Bat}} + \mathrm{n_{Inv}C_{Inv}}\right] \tag{16}$$

Similarly, for the Wind/Battery hybrid system design, the equations are modified as:

$$\mathrm{AMC} = \mathrm{n_{wt}P_{wtm}} \tag{17}$$

Whereas, the total capital cost is calculated as:

$$\mathrm{ACC} = \mathrm{CFR} \times \left[\mathrm{n_{wt}C_{wt}} + \mathrm{n_{Bat}C_{Bat}} + \mathrm{n_{Inv}C_{Inv}}\right] \tag{18}$$

In all the configurations, the number of each hybrid component is selected as the decision variables using the following boundary constraints:

$$\mathrm{n_{pv-max}} \leq \mathrm{n_{pv}} \leq \mathrm{n_{pv-min}} \tag{19}$$

$$\mathrm{n_{wt-max}} \leq \mathrm{n_{wt}} \leq \mathrm{n_{wt-min}} \tag{20}$$

$$\mathrm{n_{Bat-max}} \leq \mathrm{n_{Bat}} \leq \mathrm{n_{Bat-min}} \tag{21}$$

The number of inverters is selected as 4 and 3 for PV/Wind/Battery and the other two configurations respectively.

3.2 Constrain formulation

The optimization problem of the hybrid renewable system is to determine the right combination of PV panels, wind turbines and batteries which gives the minimum ACC and AMC. The ACC and AMC should, in turn, minimize the TAC given in Eq. (22) while satisfying the following constraints:

$$\begin{aligned} &n_{pv\_\min} \le n_{pv} \le n_{pv\_\max} \quad n_{wt\_\min} \le n_{wt} \le n_{wt\_\max} \quad n_{Bat\_\min} \le n_{Bat} \le n_{Bat\_\max} \\ &\text{and,} \\ &n_{pv}, n_{wt} and\, n_{Bat} = Integer \end{aligned} \tag{22}$$

From Eq. (10), $n_{pv\_\min}$ and $n_{pv\_\max}$ are the lower and upper limit of n_{pv}; $n_{wt\_\min}$ and $n_{wt\_\max}$ are the lower and upper bound of n_{wt} and, $n_{Bat\_\min}$ and $n_{Bat\_\max}$ are the lower and upper bound of n_{Bat}.

The charge quantity of the battery bank at any time should satisfy the following constraints

$$SOC(t_{\min}) \le SOC(t) \le SOC(t_{\max}) \tag{23}$$

The SOC whose maximum and minimum charge quantity is defined by $SOC(t_{\min})$ *and* $SOC(t_{\max})$ and is the battery State of Charge, determined by Eq. (8)

$$SOC(t) = SOC(t-1) \times (1-\omega) + \left[\frac{P_L(t)}{\eta_{Inv}} - (P_{PV}(t) - P_{WT}(t))\right] \times \eta_{BC} \tag{24}$$

where ω is the hourly self-discharge rate of the battery, η_{BC} is the battery bank discharge efficiency, η_{Inv} is the inverter efficiency.

The maximum charge quantity of the battery bank SOC (t_{max}) takes the value of the nominal capacity of the battery (S_{Bat}) and the minimum charge quantity of the battery bank SOC (t_{min}) is obtained from the maximum Depth of Discharge (DOD) as in Eq. (9).

$$SOC(t_{\min}) = (1 - DOD) \times S_{Bat} \tag{25}$$

4. Modified smell agent optimization

Studies have shown that one of the control parameters which affect the overall performance of smell agent optimization is the olfaction capacity of the agent. Since the ability of the agent to perceive a smell molecule largely depend on the size of the olfactory lobe, proper choice of olfaction capacity will influence the searching ability of Smell Agent Optimization positively. Unlike in the original SAO, where the olfaction capacity is selected arbitrarily, this research developed a model to select the olfaction capacity dynamically. This is to ensure that, the olfaction capacity changes as the algorithm iterate through the optimization process. Assuming the initial olfaction capacity assigned to an agent is given as *olf*, then the dynamic *olf* which is donated as olf_{dyn} can be calculated as

$$olf_{dyn} = olf \times e^{\left(\frac{itr}{itr_{\max}}\right)} \tag{26}$$

The dynamic olfaction capacity given in Eq. (26) is formulated such that, the perception capability of the agent exponentially as the agent approaches the object generating the smell molecules (i.e., optimal solution). The equation is used to modify the trailing mode solution search of SAO. The flow chart for the implementation of the dynamic SAO called the iSAO is given in figure below.

Note: From **Figure 2**, the part highlighted in red shows the improvement added to the standard smell agent optimization.

Step 1: The parameters required to implement the algorithm are initialized. These parameters are the population (positions) of the smell molecules, the initial velocity, search dimension, temperature, Boltzmann constant, random mode step movement and the number of iterations.

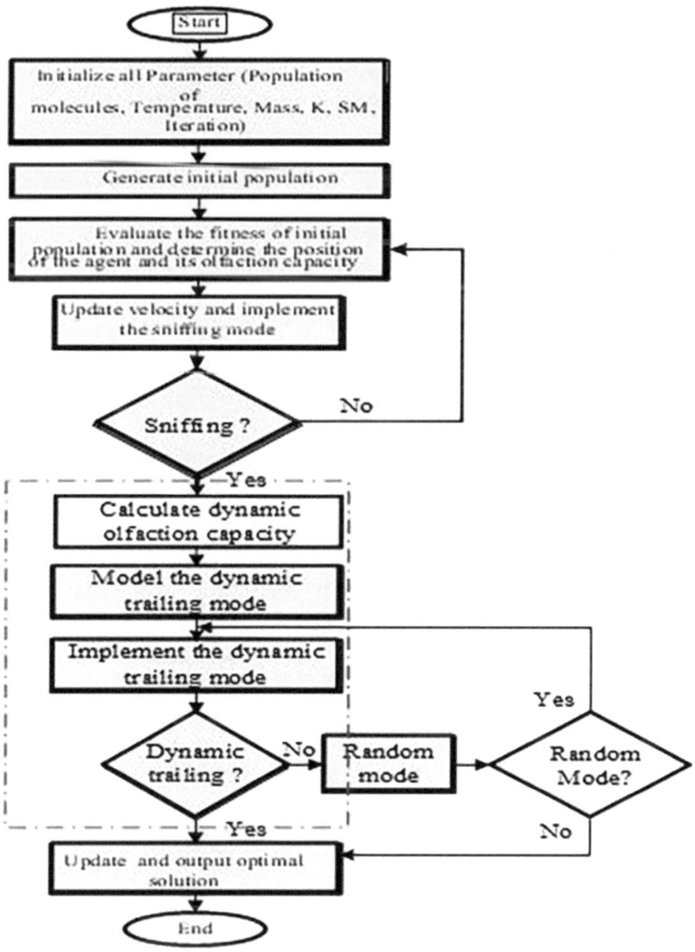

Figure 2.
Illustrates improved smell agent optimization flow chart [1].

Step 2: Randomly generate the initial position of the smell molecules in the search space and assign the velocity of each molecule.

Step 3: The fitness of the generated molecules position and velocity in step 2 are evaluated.

Step 4: The velocity of each molecule is updated.

Step 5: The fitness of the sniffing mode and position of the agent in step 2 with the position of the molecule having the best sniffing fitness are evaluated.

Step 6: The dynamic olfaction capacity is calculated, the position of the molecules with the worst sniffing fitness is determined and the trailing mode behavior is performed.

Step 7: The fitness of the trailing mode is evaluated.

Step 8: The fitness of the trailing mode with the fitness obtained during the fitness mode are compared.

Step 9: If the trailing mode fitness is better than the sniffing mode fitness, step 6 is repeated to step 8 until the smell source is determined. If the sniffing mode fitness is better than the trailing mode fitness, then move to step 10.

Step 10: Random mode behavior is performed.

Step 11: The fitness of the random mode is evaluated.

Step 12: If the fitness of the random mode is better than the fitness of the trailing mode, then determine the new random position of the agent and the worst random position of the molecule and perform the trailing mode again, otherwise repeat step 4 to step 9.

Step 13: Terminate if the stopping condition is satisfied else repeat step 1 to step 12 until the stopping criteria are met.

5. Important assumptions

Highlighted below are some of the important assumptions adopted for the development of the proposed iSAO;

i. The smell molecules evaporate from the smell source constantly in the direction of the agent until the smell object is found.

ii. The velocity of the object evaporating the smell molecules is negligible compared to the velocity of the smell agent. In other words, the object originating the smell is in a fixed position and cannot move.

iii. The smell source could be more than one and each source evaporate the same number of smell molecules with varying concentration [13].

iv. The olfaction capacity is dynamic in nature.

6. Overview of data collection

The data from **Table 1** were compiled from first a questionnaire distributed around Abuja environs to understand the energy behavior of the Nigerian household

Solar (W/m^2)	Wind speed (m/s)	Load (kW)
0	22	6.4
0	23	6.2
0	25	5.8
0	26	5.8
0	26	6
0	26	7.3
198	26	9.1
562	24	12.5
830	22	13.4
800	20	13.4
811	19	13.2
813	17	13.8
403	15	13.2
803	16	13.7
844	16	13.2
678	17	14
322	12	16.3
0	7	18.9
0	3	22.2
0	4	17.9
0	5	12.5
0	6	9.1
0	13	7.5
0	19	6.8

Table 1.
Data collected for the project.

and from data collected by Sunergy, the renewable energy department of Anjeed Innova group.

Table 1 is the data for the daily wind speed, solar insolation and the required daily load. The load peaks at 22.2 kW with a minimum of 5.8 kW.

The graph of **Figure 3** shows a plot of the load required in kW on the y-axis against and the hours of the day. The load steadily climbs from about 6 am at 7.3 kW and maintains by 9 am at about 13 kW then starts to rise again at 5 pm to 16.3 kW then peaks at 22.3 kW by 7 pm then drops till around 6 kW till the next day. This is typical of the average Nigerian family activity.

The graph **Figure 4** below shows a plot of wind speed in m/s and the solar insolation in W/m^2 on the y-axis against the hours of the day on the x-axis. Because of the intermittent nature of solar and wind energy, sunlight is to be expected for about 12 hours and the wind speed rises and falls rapidly throughout the day.

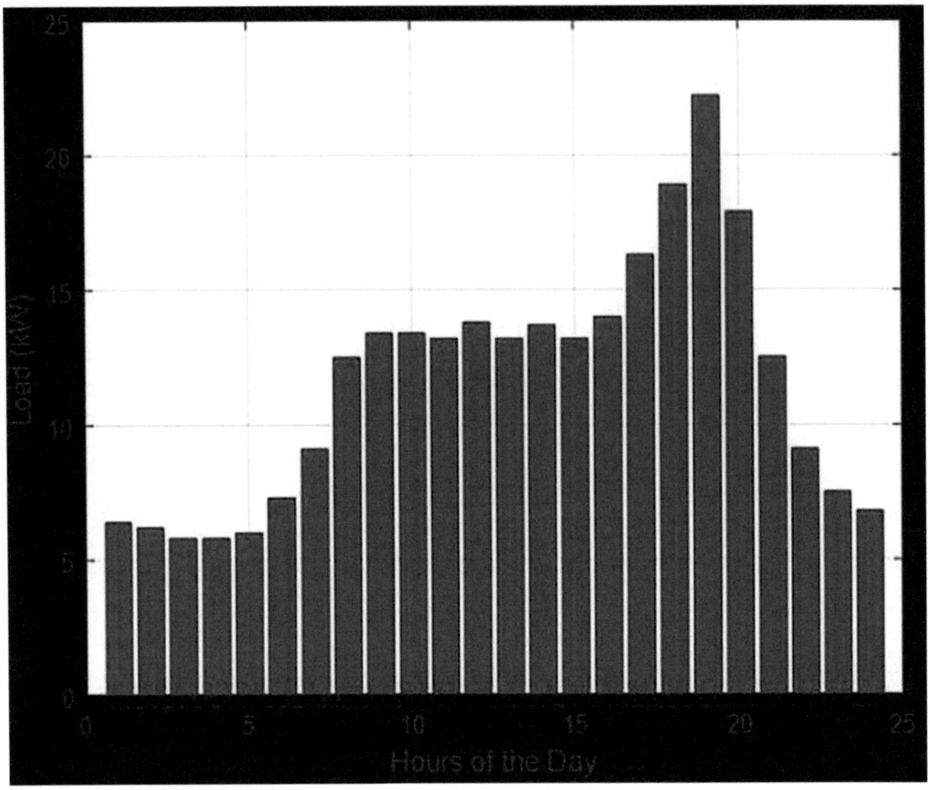

Figure 3.
Illustrating hourly load requirement [1].

7. Results and discussion

The algorithms attempt to find the optimum number of PV panels, wind turbines, and batteries (NBatt) in PV/WT/battery, PV/battery and WT/battery configurations. The minimum and maximum numbers of each component are set to 0 and 100 for the solar PV and wind turbines, respectively the 50 for the battery storage.

Also, the olfaction capacity is dynamic with a total of 100 iterations and to calculate the accuracy of results, 10 independent runs are performed and the results are reported below.

8. Analysis

i. Standard smell agent optimization

Table 2 illustrates the iterations 7 for the PV, wind turbine and battery and from the results, the algorithm finds the best and the worst combination economically with optimum reliability.

This program was run on Matlab R2017a and the graph is plotted with the annual cost on the y-axis with the iteration number on the x-axis (**Figure 5**).

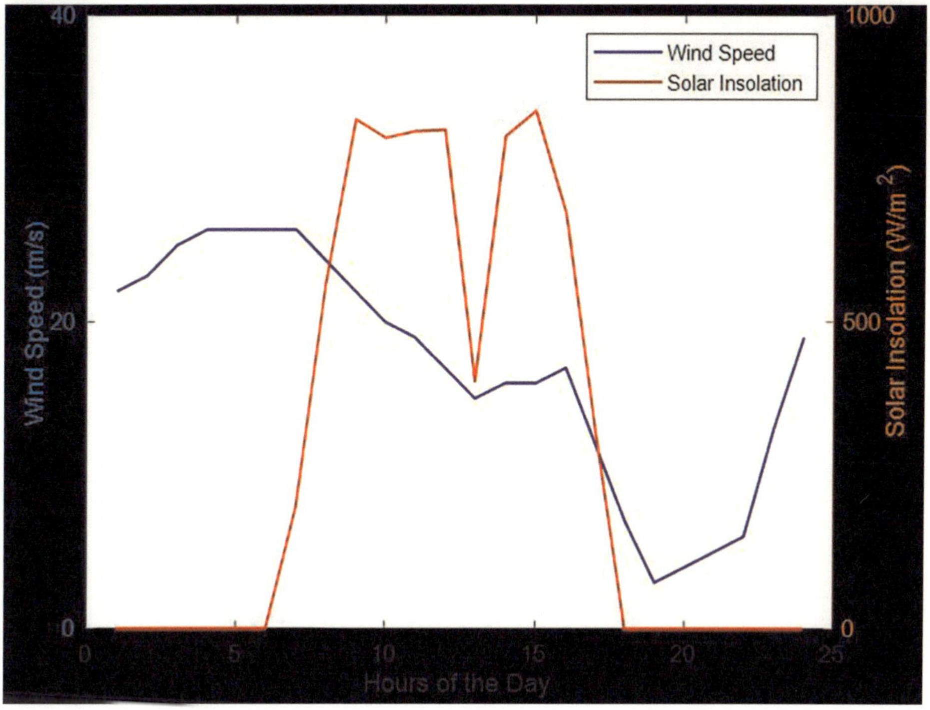

Figure 4.
Illustrating daily wind speed and solar insolation [1].

Table 3 WT and battery using the improved smell agent optimization. and battery. From the results, the algorithm finds the best and the worst combination economically with optimum reliability and the graph is plotted with the annual cost on the y-axis with the iteration number on the x-axis (**Figure 6**).

Table 4 illustrates the iterations for the wind turbine and battery. From the results, the algorithm finds the best and the worst combination economically with optimum reliability and the graph is plotted with the annual cost on the y-axis with the iteration number on the x-axis (**Figure 7**).

ii. Improved smell agent optimization

Table 5 illustrates the iterations for the PV/WT and battery using the improved smell agent optimization. From the results, the algorithm finds the best and the worst combination economically with optimum reliability and the graph is plotted with the annual cost on the y-axis with the iteration number on the x-axis (**Figure 8**).

Table 6 illustrates the iterations for the PV and battery using the improved smell agent optimization. From the results, the algorithm finds the best and the worst combination economically with optimum reliability and the graph is plotted with the annual cost on the y-axis with the iteration number on the x-axis (**Figure 9**).

Table 7 illustrates the iterations for the WT and battery using the improved smell agent optimization. From the results, the algorithm finds the best and the

S/no	Npv	Nwt	Nbat	pvCost	CwtCost	Bat_Cost	Conv_Cost	Total_Cost	LCE	LpSP	E_Ex
1	16	34	50	1.28E+04	5.59E+03	3.23E+02	1.07E+02	2.04E+04	0.039754	0.173659	93.64366
2	80	61	8	6.42E+04	1.00E+04	51.62501	1.07E+02	7.59E+04	0.02212	0.173659	1.68E+02
3	0	0	3	0	0	19.35938	1.07E+02	1.63E+03	Inf	0.187702	-0.01159
4	92	100	12	7.38E+04	1.65E+04	77.43751	1.07E+02	9.19E+04	0.013504	0.173659	2.76E+02
5	25	71	50	2.01E+04	1.17E+04	3.23E+02	1.07E+02	3.37E+04	0.019042	0.173659	1.96E+02
6	47	100	38	3.77E+04	1.65E+04	2.45E+02	1.07E+02	5.60E+04	0.013517	0.173659	2.75E+02
7	61	82	26	4.89E+04	1.35E+04	1.68E+02	1.07E+02	6.42E+04	0.016474	0.173659	2.26E+02
8	48	75	17	3.85E+04	1.23E+04	1.10E+02	1.07E+02	5.26E+04	0.018016	0.173659	2.07E+02
9	95	68	50	7.62E+04	1.12E+04	3.23E+02	1.07E+02	8.93E+04	0.019839	0.173659	1.88E+02
10	55	100	11	4.41E+04	1.65E+04	70.98439	1.07E+02	6.23E+04	0.013514	0.173659	2.75E+02

Table 2.
Summary of the results for the hybrid systems obtained by SAO algorithm for 10 runs of the PV/WT/BAT configurations.

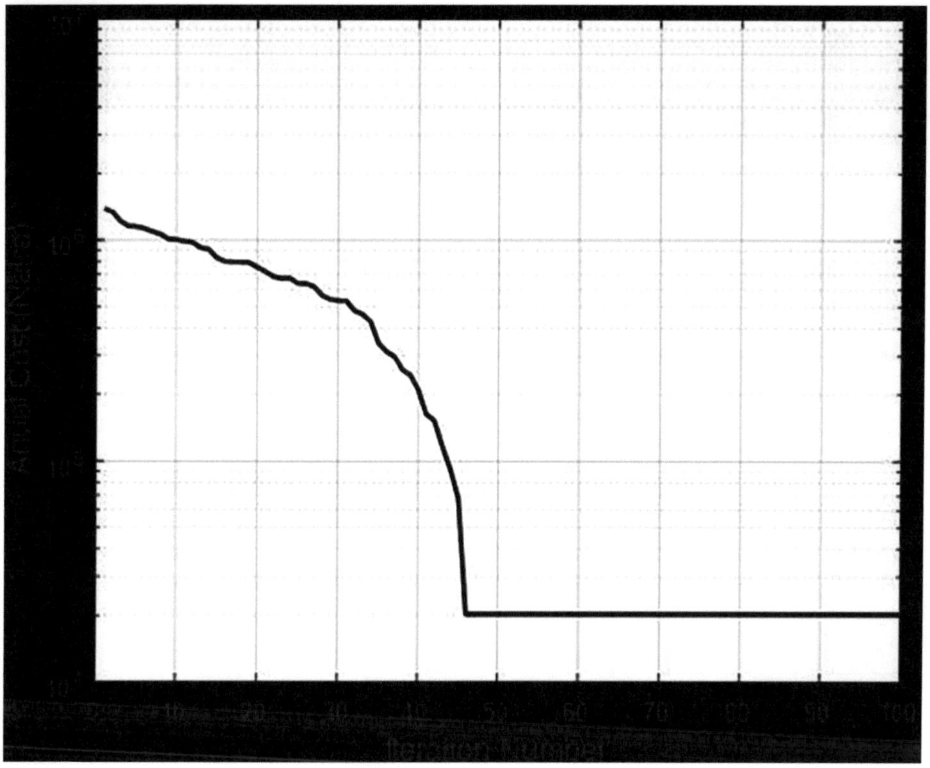

Figure 5.
Illustration of annual cost against iteration number for PV/wt/battery configurations [1].

S/no	Npv	Nbat	pvCost	Bat_Cost	Conv_Cost	Total_Cost	LCE	LpSP	E_Ex
1	95	18	7.62E+04	1.16E+02	8.00E+01	7.75E+04	6.851904	0.173659	0.531792
2	67	18	5.37E+04	1.16E+02	8.00E+01	5.51E+04	9.715386	0.173659	0.371637
3	45	35	3.61E+04	2.26E+02	8.00E+01	3.75E+04	14.46513	0.173659	0.2458
4	57	50	4.57E+04	3.23E+02	8.00E+01	4.73E+04	11.41984	0.173659	0.314438
5	90	16	7.22E+04	1.03E+02	8.00E+01	7.35E+04	7.232565	0.173659	0.503192
6	99	33	7.94E+04	2.13E+02	8.00E+01	8.08E+04	6.575059	0.173659	0.554671
7	88	0	7.06E+04	0	8.00E+01	7.18E+04	7.396941	0.173659	0.491753
8	43	50	3.45E+04	3.23E+02	8.00E+01	3.60E+04	15.13793	0.173659	0.234361
9	78	0	6.26E+04	0	8.00E+01	6.38E+04	8.345267	0.173659	0.434555
10	84	0	6.74E+04	0.00E+00	8.00E+01	6.86E+04	7.749177	0.173659	0.468874

Table 3.
Summary of the results for the hybrid systems obtained by SAO algorithm for 10 runs of the PV/BAT configurations.

worst combination economically with optimum reliability and the graph is plotted with the annual cost on the y-axis with the iteration number on the x-axis (**Figure 10**).

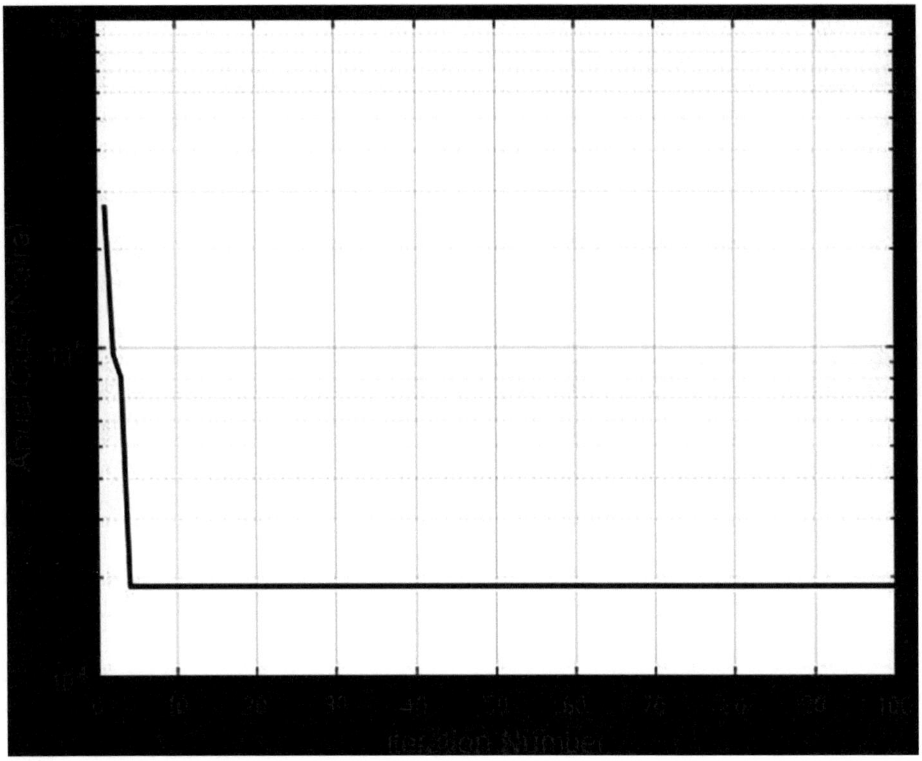

Figure 6.
Illustration of annual cost against iteration number for PV/battery configurations [1].

S/no	Nwt	Nbat	CwtCost	Bat_Cost	Conv_Cost	Total_Cost	LCE	LpSP	E_Ex
1	29	17	4.77E+03	1.10E+02	8.00E+01	6.09E+03	0.046654	0.093959	7.98E+01
2	61	28	1.00E+04	1.81E+02	8.00E+01	1.14E+04	0.02218	0.093959	1.68E+02
3	100	30	1.65E+04	1.94E+02	8.00E+01	1.79E+04	0.01353	0.093959	2.75E+02
4	16	22	2.63E+03	1.42E+02	8.00E+01	3.98E+03	0.084561	0.093959	4.40E+01
5	30	28	4.94E+03	1.81E+02	8.00E+01	6.32E+03	0.045099	0.093959	8.25E+01
6	94	32	1.55E+04	2.07E+02	8.00E+01	1.69E+04	0.014393	0.093959	2.59E+02
7	100	50	1.65E+04	3.23E+02	8.00E+01	1.80E+04	0.01353	0.093959	2.75E+02
8	17	50	2.80E+03	3.23E+02	8.00E+01	4.33E+03	0.079587	0.093959	4.68E+01
9	53	29	8.72E+03	1.87E+02	8.00E+01	1.01E+04	0.025528	0.093959	1.46E+02
10	33	16	5.43E+03	1.03E+02	8.00E+01	6.74E+03	0.040999	0.093959	9.08E+01

Table 4.
Summary of the results for the hybrid systems obtained by SAO algorithm for 10 runs of the WT/BAT configurations.

Figure 7.
Illustration of annual cost against iteration number for wt/battery configurations [1].

9. Comparative analysis

It should be noted from the table that only the best results obtained by each algorithm are used as a metric for comparing the performance of the algorithm. The algorithm is developed by the phenomenon of smell and the trailing behavior of agents in identifying smell sources. The results showed that the improved SAO is efficient and can compete with other computational intelligent algorithms.

Table 8 illustrating the best, average and the standard deviation values for the total annual cost, Levelized cost of energy, loss of power supply probability and excess energy of the standard SAO PV/WT/batt configurations.

Table 9 illustrating the best, average and the standard deviation values for the total annual cost, Levelized cost of energy, loss of power supply probability and excess energy of the standard SAO PV/batt configurations.

Table 10 illustrating the best, average and the standard deviation values for the total annual cost, Levelized cost of energy, loss of power supply probability and excess energy of the standard SAO WT/batt configurations.

Table 11 illustrating the best, average and the standard deviation values for the total annual cost, Levelized cost of energy, loss of power supply probability and excess energy of the improved SAO PV/WT/batt configuration.

S/no	Npv	Nwt	Nbat	pvCost	CwtCost	Bat_Cost	Conv_Cost	Total_Cost	LCE	LpSP	E_Ex
1	0	45	41	0	7.40E+03	2.65E+02	1.07E+02	9.28E+03	0.030066	0.093959	1.24E+02
2	63	100	28	5.05E+04	1.65E+04	1.81E+02	1.07E+02	6.88E+04	0.013512	0.173659	2.76E+02
3	92	36	48	7.38E+04	5.92E+03	3.10E+02	1.07E+02	8.16E+04	0.037384	0.173659	99.58211
4	95	100	22	7.62E+04	1.65E+04	1.42E+02	1.07E+02	9.44E+04	0.013503	0.173659	2.76E+02
5	72	100	41	5.78E+04	1.65E+04	2.65E+02	1.07E+02	7.61E+04	0.01351	0.173659	2.76E+02
6	36	64	50	2.89E+04	1.05E+04	3.23E+02	1.07E+02	4.13E+04	0.021116	0.173659	1.76E+02
7	14	0	15	1.12E+04	0	96.79689	1.07E+02	1.29E+04	46.49506	0.173659	0.068486
8	31	92	50	2.49E+04	1.51E+04	3.23E+02	1.07E+02	4.19E+04	0.014696	0.173659	2.53E+02
9	17	88	22	1.36E+04	1.45E+04	1.42E+02	1.07E+02	2.99E+04	0.015369	0.173659	2.42E+02
10	100	3	50	8.02E+04	4.94E+02	3.23E+02	1.07E+02	8.26E+04	0.421769	0.173659	8.816014

Table 5.
Summary of the results for the hybrid systems obtained by improved smell agent algorithm for 10 runs of the PV/WT/BAT configurations.

Figure 8.
Illustration of annual cost against iteration number for PV/wt/battery configurations using the improved smell agent optimization [1].

S/no	Npv	Nbat	pvCost	Bat_Cost	Conv_Cost	Total_Cost	LCE	LpSP	E_Ex
1	19	17	1.52E+04	1.10E+02	1.07E+02	1.70E+04	34.25952	34.25952	0.097085
2	78	35	6.26E+04	2.26E+02	1.07E+02	6.44E+04	8.345267	0.173659	0.434555
3	32	46	2.57E+04	2.97E+02	1.07E+02	2.76E+04	20.34159	0.173659	0.171443
4	19	33	1.52E+04	2.13E+02	1.07E+02	1.71E+04	34.25952	0.173659	0.097085
5	36	30	2.89E+04	1.94E+02	1.07E+02	3.07E+04	18.08141	0.173659	0.194322
6	33	40	2.65E+04	2.58E+02	1.07E+02	2.83E+04	19.72518	0.173659	0.177163
7	100	47	8.02E+04	3.03E+02	1.07E+02	8.21E+04	6.509308	0.173659	0.560391
8	100	46	8.02E+04	2.97E+02	1.07E+02	8.21E+04	6.509308	0.173659	0.560391
9	100	29	8.02E+04	1.87E+02	1.07E+02	8.20E+04	6.509308	0.173659	0.560391
10	34	13	2.73E+04	83.89064	1.07E+02	2.90E+04	19.14502	0.173659	0.182882

Table 6.
Summary of the results for the hybrid systems obtained by improved smell agent algorithm for 10 runs of the PVBAT configurations.

Table 12 illustrating the best, average and the standard deviation values for the total annual cost, Levelized cost of energy, loss of power supply probability and excess energy of the improved SAO PV/batt configurations.

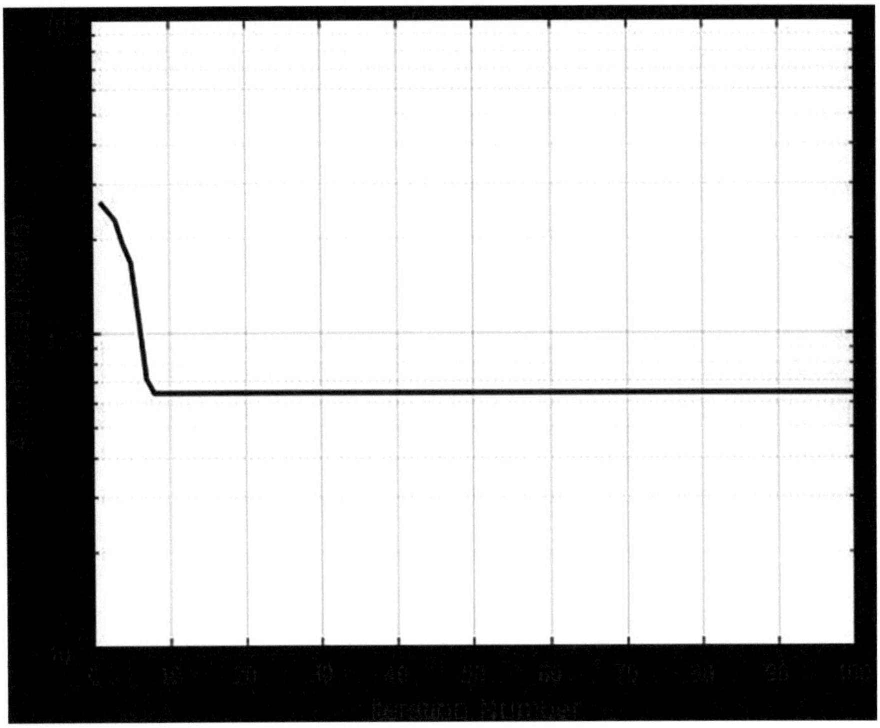

Figure 9.
Illustration of annual cost against iteration number for PV/battery configurations using the improved smell agent optimization [1].

S/no	Nwt	Nbat	CwtCost	Bat_Cost	Conv_Cost	Total_Cost	LCE	LpSP	E_Ex
1	56	29	9.21E+03	1.87E+02	1.07E+02	1.10E+04	0.02416	0.093959	1.54E+02
2	44	3	7.24E+03	19.35938	1.07E+02	8.87E+03	8.87E+03	0.093959	1.21E+02
3	1	14	1.65E+02	90.34376	1.07E+02	1.86E+03	1.352972	0.093959	2.740283
4	0	16	0	1.03E+02	1.07E+02	1.71E+03	Inf	0.187702	-0.01159
5	12	3.10E+01	1.97E+03	2.00E+02	1.07E+02	3.78E+03	0.112748	0.093959	33.0109
6	8	50	1.32E+03	3.23E+02	1.07E+02	3.25E+03	0.169122	0.093959	22.0034
7	99	28	1.63E+04	1.81E+02	1.07E+02	1.81E+04	0.013666	0.093959	2.72E+02
8	14	26	2.30E+03	1.68E+02	1.07E+02	4.08E+03	0.096641	0.093959	38.51465
9	100	15	1.65E+04	96.79689	1.07E+02	1.82E+04	0.01353	0.093959	2.75E+02
10	100	50	1.65E+04	3.23E+02	1.07E+02	1.84E+04	0.01353	0.093959	2.75E+02

Table 7.
Summary of the results for the hybrid systems obtained by improved smell agent algorithm for 10 runs of the WT/BAT configurations.

Table 13 illustrating the best, average and the standard deviation values for the total annual cost, Levelized cost of energy, loss of power supply probability and excess energy of the improved SAO WT/batt configurations.

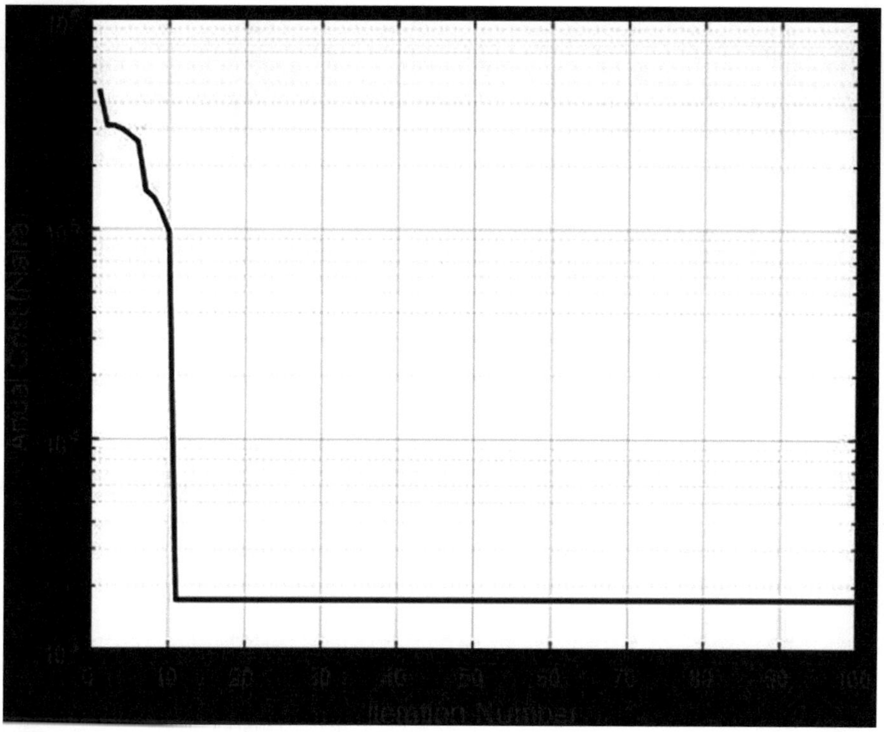

Figure 10.
Illustration of annual cost against iteration number for wt/battery configurations using the improved smell agent optimization [1].

	Total annual cost	LCE	LPSP	Ex.Energy
Best	1.63E+03	0.013504	0.175063	−0.01159
Avg	5.48E+04	0.017578	0.175063	190.4398
SD	2.91E+04	0.009886	0.004441	87.82577

Table 8.
Illustrating the best, average and the standard deviation values for the total annual cost, Levelized cost of energy, loss of power supply probability and excess energy of the standard SAO PV/WT/batt configurations.

	Total annual cost	LCE	LPSP	Ex.Energy
Best	3.75E+04	6.575059	0.173659	0.234361
Avg	6.12E+04	9.488919	0.173659	0.415107
SD	1.63E+04	3.157008	2.92569E-17	0.117073

Table 9.
Illustrating the best, average and the standard deviation values for the total annual cost, Levelized cost of energy, loss of power supply probability and excess energy of the standard SAO PV/batt configurations.

	Total annual cost	LCE	LPSP	Ex.Energy
Best	1.01E+04	0.01353	0.093959	1.46E+02
Avg	1.02E+04	0.038606	0.093959	1.47E+02
SD	5.60E+03	0.026223	0	9.32E+01

Table 10.
Illustrating the best, average and the standard deviation values for the total annual cost, Levelized cost of energy, loss of power supply probability and excess energy of the standard SAO WT/batt configurations.

	Total annual cost	LCE	LPSP	Ex.Energy
Best	1.29E+04	0.013503	0.093959	0.068486
Avg	5.39E+04	4.707598	0.165689	1.73E+02
SD	3.07E+04	14.68316	0.025204	1.09E+02

Table 11.
Illustrating the best, average and the standard deviation values for the total annual cost, Levelized cost of energy, loss of power supply probability and excess energy of the improved SAO PV/WT/batt configuration.

	Total annual cost	LCE	LPSP	Ex.Energy
Best	1.70E+04	6.509308	0.173659	0.171443
Avg	4.60E+04	17.36854	3.582245	0.303571
SD	2.81E+04	10.66591	10.77889	0.200066

Table 12.
Illustrating the best, average and the standard deviation values for the total annual cost, Levelized cost of energy, loss of power supply probability and excess energy of the improved SAO PV/batt configurations.

	Total annual cost	LCE	LPSP	Ex.Energy
Best	1.10E+04	0.01353	0.1877	−0.1159
Avg	8.92E+03	886.9933	0.103333	1.19E+02
SD	7.05E+03	2804.288	0.029644	1.18E+02

Table 13.
Illustrating the best, average and the standard deviation values for the total annual cost, Levelized cost of energy, loss of power supply probability and excess energy of the improved SAO WT/batt configurations.

10. Findings of the study

The findings of this study are as follows:

i. The percentage of improvement for the total annual cost and loss of power supply probability between the standard smell agent optimization and the improved smell agent optimization is 79% and 53.4% respectively.

ii. The sizing technique is suitable for off-grid and grid configurations.

iii. The technique can be used for various component configurations.

iv. The improved SAO is better at detecting and discriminating scent agents in the environment due to varying olfactory capacity.

v. The improved SAO has a better capacity to follow a scent or odor plum.

11. Conclusions

Unlike in the original SAO, where the olfaction capacity is selected arbitrarily, this research developed a model to select the olfaction capacity dynamically. This is to ensure that, the olfaction capacity changes as the algorithm iterate through the optimization process.

This simple modification improved the trailing and tracking to obtain the cost effective HRES design.

12. Recommendations

The improved SAO has both constant and dynamic variables which play a significant role in the general performance of the algorithm. For this reason, the following areas are highlighted for consideration.

i. Practically, an increase in the temperature of gas molecules increases its evaporation and the velocity of the gas. In this study, these values are constant and since smell molecules in SAO are considered as gas molecules, a method to adaptively select temperature can be considered that the temperature has a decreasing value as the algorithm moves towards the optimum solution. This will enable the algorithm to converge faster and eventually terminate the process when the minimum value of the temperature is attained.

ii. In this work, all gas molecules are assumed a fixed mass. That is not always the case so making the mass of the gas molecules adaptive can be considered. E.g., larger values of mass favor the exploitation capability of the agent while the smaller value will favor exploration capability.

iii. This suggests the possibility of developing a novel algorithm using other sensory systems such as sense of taste, sense of feel and sense of hearing. It is logical to suggest that an algorithm can be developed using other senses like the use of the senses of smell and taste coordinated through the chemo-sensation process.

iv. The improved SAO can be hybridized or cascaded with similar computational intelligent algorithms for improved performance.

v. The improved SAO can be applied to problems related to other fields like image and signal processing, power systems and sensor networks etc.

13. Limitations

The aim of the research is the development of an improved smell agent optimization sizing technique algorithm for a hybrid renewable energy system for off-grid use to obtain the most cost effective HRES design. This was achieved by modifying the smell agent optimization technique and this has been successfully achieved. However, it could not establish that all the molecules evaporating from a smell source are accounted for by the agent. It is assumed that the agent only makes its decision on the smell molecules it perceived.

Author details

Akawu Shekari Biliyok[1*] and Salawudeen Ahmed Tijani[2]

1 Department of Elect/Elect Engineering, Nile University of Nigeria, Abuja, Nigeria

2 Department of Elect/Elect Engineering, University of Jos, Nigeria

*Address all correspondence to: scottyonline36@gmail.com

References

[1] Biliyok AS. Developing an Improved Smell Agent Optimization Sizing Technique for a Grid-Independent Hybrid Renewable Energy System [thesis]. Abuja, Nigeria: Department of Electrical Electronic Engineering, Nile University of Nigeria, Abuja; 2021

[2] Wikipedia. Hybrid Renewable Energy System. 2020. Available from: https://en.wikipedia.org/wiki/Hybrid_renewable_energy_system [Accessed: 09 November 2020]

[3] Chung CJ. Knowledge-Based Approaches to Self-Adaptation in Cultural Algorithms [thesis]. MI, USA: Wayne State University; 1997

[4] Baba Yachilla Alhaji. Development of an Intelligent Pid Controller Using an Improved Artificial Fish Swarm Optimization Algorithm for the Control of DC-Motor [thesis]. Abuja, Nigeria: Department of Computer Engineering, Nile University of Nigeria, Abuja; 2016

[5] Chedid R, Sawwas A. Optimal placement and sizing of photovoltaics and battery storage in distribution network. Energy Storage. 2019;**1**:e46

[6] Maleki A, Khajeh MG, Ameri M. Optimal sizing of a grid-independent hybrid renewable energy system incorporating resource uncertainty, and load uncertainty. International Journal of Electrical Power and Energy Systems. 2016;**83**:514-524

[7] Vinod Chandra SS. Smell detection agent-based optimization algorithm. Journal of the Institution of Engineers (India). 2016;**97**(Issue 3):431-436

[8] Iwayemi A. Nigeria's Dual Energy Problems: Policy Issues and Challenges. International Association for Energy Economics; 2008

[9] Sambaiah KS, Jayabarathi T. Loss minimization techniques for optimal operation and planning of distribution systems: A review of different methodologies. International Transactions on Electrical Energy Systems. 2019;**30**(Issue 2):e12230

[10] Okanlawon L. The Potential of Nigeria's Residential Solar Rooftop Systems. 2015. Available from: https://www.renewableenergyworld.com/storage/the-potential-of-nigerias-residential-solar-rooftop-systems/#gref [Accessed: 04 March 2021]

[11] Salawudeen AT, Mu'azu MB, Yusuf A, Adedokun AE. A novel smell agent optimization (SAO): An extensive CEC study and engineering application. Knowledge-Based Systems. 2021;**232**:107486

[12] Salawudeen AT, Mu'azu MB, Yusuf A, Adedokun AE. From smell phenomenon to smell agent optimization (SAO): A feasibility study. In: Proceedings of ICGET. 2018

[13] Salawudeen AT. Development of a Smell Agent Optimization Algorithm for Combinatorial Optimization Problems. Zaria, Nigeria: Department of Computer Engineering, Faculty of Engineering, ABU Zaria; 2018

Chapter 4

Use of Induction Generators in Small Hydro Power Generation System Feeding Isolated Load in Remote Mountainous Regions of Himalayas

Umesh C. Rathore and Sanjeev Singh

Abstract

Providing reliable and clean power from conventional grid in remote mountainous regions is always a challenging task due to tough geographical and climatic conditions. Renewable energy sources based power plants such as small hydro power plants play a significant role in meeting the power requirements in these remote locations in mountainous regions. Synchronous generators are the most commonly used generators in small hydro power plants. However, with the advancement in controller technology for voltage and frequency control, induction generators are nowadays preferred in renewable energy conversion systems. Self excited induction generators (SEIG) in small hydro power plants feeding isolated domestic loads are more suitable due to their inherent advantages as compared to conventional synchronous generators. This chapter deals with the usefulness of electronic load controller used in voltage and frequency control of self-excited induction generator used in small hydro power plant feeding isolated load in remote mountainous regions of Himalayas.

Keywords: self-excited induction generator, small hydro, isolated load, electronic load controller, dump load

1. Introduction

To meet electricity demand of people living in mountainous terrain is always a challenging task. Due to adverse geographical conditions, the cost of laying and maintaining electrical power distribution network increases. Under such conditions, using renewable energy sources is the best option to meet their energy requirement. It will also cause less adverse impact on environment as compared to impact caused by using conventional energy sources. Advancement in technology and growing concern on environmental issues worldwide have motivated and compelled the world to maximize the use of renewable sources of energy such as wind, small hydro, solar, geothermal, tidal, biomass, ocean thermal etc. [1]. These renewable energy sources are best choice to compliments the conventional sources of energy and also to meet the growing energy demand of the people living in remote areas where use of conventional grid for power distribution is difficult. An estimated 20–25% of the global final energy consumption comes from various renewable energy sources and is increasing day by day.

Hydroelectric power is the most appropriate, cost effective and environment friendly renewable energy source suitable for providing electrical power in remote mountainous regions which actually have the huge hydro power potential for electricity generation. Especially small hydro power plants play a significant role in meeting the electricity requirement of people living in remote mountainous regions in isolated mode. In comparison to large hydro power plants, micro/pico hydro power plants are environment friendly; require less investment and easy operating with minimal impact on environment. An estimated electric power potential of more than 200,000 MW from small hydro exists worldwide, out of which around 50,000 MW has been tapped so far. India also has an estimated potential of more than 5000 MW capacity of small hydro identified through more than 2000 sites in various states of India mostly located in Himalayan region. In hydro power plants, the erratic rainfall affects the water discharge in catchment areas which further affects the generated kWh output. Most of the small hydro power plants are located in mountainous terrains. However, due to the effect of global warming on overall environment around the globe and erratic rainfall due to disturbance in environment, the output of these small hydro-electric power plants (SHEPP) is not consistent. Similar varying water input conditions exist almost in the entire mountainous region as the water flow rate depends upon the average rainfall and quantity of snow on the mountain peaks in concerned catchment area which feed the small hydro power plants. Annual power generation trend based on the data collected from a cluster of small hydro power plants in Himalayan region shows the variation in power generation against the rated capacity due to varying inputs as shown in **Figure 1** [2].

Another crucial factor in deciding the appropriate technology for isolated type of pico-hydro power plants is the type of loading on the power plant. Generally the load requirement in remote locations varies as the load fluctuates as per the requirement.

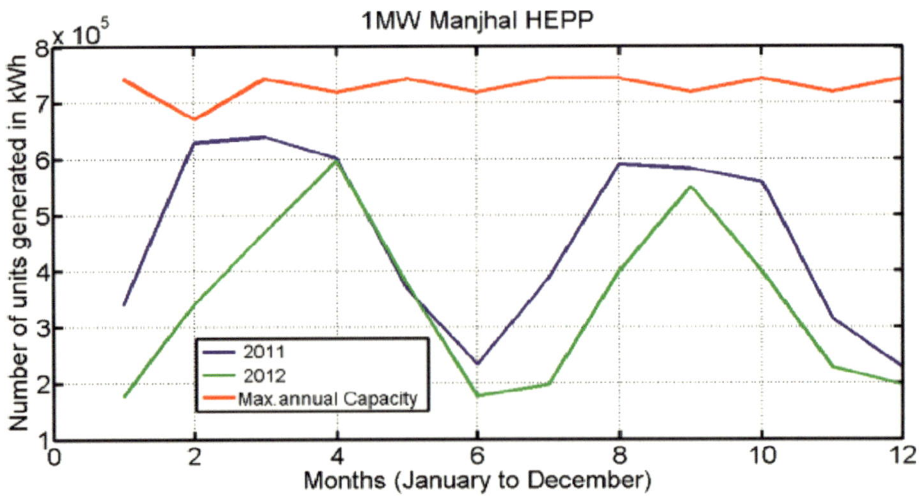

Figure 1.
Annual kWh generations in 1 MW Manjhal HEPP.

Daily load requirement varies during 24 hours in remote areas as most of the electrical load comprises of lighting and heating. Electrical loading profile in the rural remote areas can be best generated using the data from any 11 kV feeder feeding these regions. **Figures 2** and **3** show the electrical loading pattern as the per logged data collected from various 11 kV feeders feeding the rural villages in District Chamba, Himachal Pradesh situated in the Himalayan mountainous region during winter and summer months of 2013 and 2015 [2, 3].

Domestic electrical load in mountainous regions generally comprises of lighting and heating load. Apart from this there is few small industrial load too operate mostly in day times. Electrical loading pattern as shown in **Figures 2** and **3**, indicate that during the evening time between 6:30 PM to 10 PM, electrical load is maximum with peak loading at around 8 to 9 PM. The load is minimal during mid night to 6 AM and which is about 20% of the peak load. This loading pattern is of much importance when the rural isolated load is fed from a small hydro power plant without conventional grid and operating in isolated mode. Under these conditions, the use of appropriate power conversion technologies such as use of asynchronous generators (induction generators) [1] is useful due to their adaptability to load variation and to keep voltage and frequency within permissible limits.

Recent advancement in technology and widespread use of advanced computer based controllers in renewable energy conversion systems for power generation, have increased the use of induction generators which offer increased efficiency, better control and better co-ordination with grid based power supply systems or in isolated modes [1]. For isolated load, pico-hydro power plants involving self-exciting induction generators are cost effective and reliable and are best suited under varying input and fluctuating load conditions in mountainous remote locations. In these isolated pico-hydro power plants during off-load periods, the generated power can be used for other applications such as flour mills, water pumping [3], and heating of domestic homes etc. As the effect of water discharge variation is less on pico-hydro power plants, these small hydro power plants use uncontrolled hydro-turbines due to continuous availability of required water discharge. These small hydro power plants

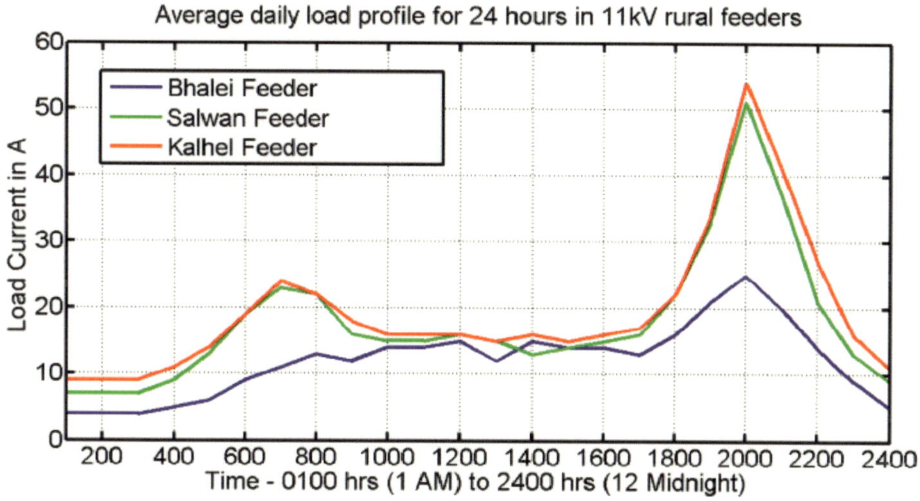

Figure 2.
Electrical loading pattern in remote rural areas in hilly regions.

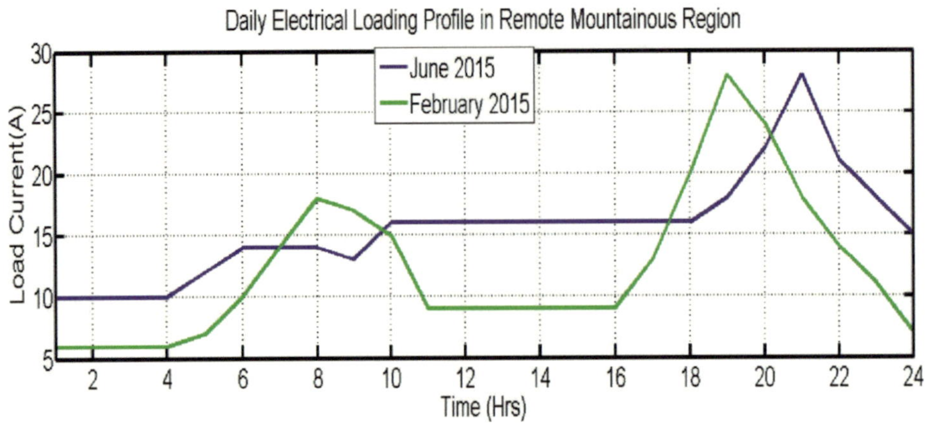

Figure 3.
Electrical loading profile of mountainous regions.

are installed in the remotest part of the mountainous region feeding isolated load for small populated areas having small electrical load requirement. Isolated Pico hydro power plants are useful in meeting the electricity requirements of these remote locations where supplying electrical power using conventional grid is not feasible due to tough climatic and geographical conditions.

This chapter presents the suitability of self-excited induction generators feeding isolated load in remote locations. After evaluating the performance characteristics of self-excited induction generators using MATLAB/Simulink based models, the simulated results are validated experimentally. The setup comprises of 2.2 kW, 4 pole, 415 V, 3-phase, squirrel cage induction motor operated as 3- phase self-excited induction generator (SEIG) driven by a 3.7 kW rated DC shunt motor which is acting as prime-mover. For self-excitation of SEIG, a 3-phase capacitor bank with different required ratings of capacitors in star and delta mode has been used. Experimental electrical load comprises of a 3-phase lamp load and a single/three-phase inductive load. The performance of SEIG is evaluated at no load and at different loads (balanced & un-balanced load) emulating the actual electrical load conditions in remote mountainous regions.

2. Induction generators

Induction generators are most suitable for renewable energy conversion systems involving wind & small hydro. The difference between induction motor and induction generator lies in their rotor speed. The rotor speed is more than the synchronous speed in induction generator and in induction motor the rotor speed is slightly less than the synchronous speed of the induction machine. In induction generators, reactive power is consumed rather than supplying from it. Induction generator supplies only the real power (kW) to the system to which it is connected. The kVAr required by the induction generator and loads on the system must be supplied from separate source such as capacitor banks. The induction generators are classified [4–6] as Self Excited Induction Generator (SEIG), Doubly Fed Induction Generator (DFIG) and Permanent Magnet Induction Generator (PMIG).

2.1 Self-excited induction generator

Self-excited induction generator is suitable for micro or pico hydro power plants feeding an isolated load. In stand-alone mode of operation of SEIG, the reactive power must be fed externally using capacitors as shown in **Figure 4** to establish the required magnetic field. Capacitors banks are connected in parallel with the induction generator terminals. These capacitor banks are connected in either star or delta modes. When the speed of induction machine rotor exceeds the synchronous speed, the residual magnetic field in the rotor starts generating a voltage across the SEIG terminals and which is further supplemented by the capacitor current to strengthen or reinforce the magnetic field and system builds up voltage with an increasing excitation.

As the speed increases, the capacitor impedance further decreases and excitation increases. This further results in increase in generated voltage of the induction generator till the voltage is limited by saturation of magnetic circuit in the induction machine. For voltage build up, there must be some residual magnetism present in the rotor of induction generator. The required value of capacitance depends upon the actual value of kVAr required and there are various techniques to calculate the exact required value of capacitance in SEIG system [7–10]. The minimum capacitance required is inversely proportional to square of the speed of rotation and to the peak value of saturated magnetizing reactance. The value of capacitance is also affected by load impedance and power factor. The maximum power output from the isolated SEIG depends upon its terminal capacitance and the speed of the generator. There must be a threshold speed called cut-off speed below which no excitation is possible at any capacitor value.

2.2 Control mechanism in SEIG

The main aim of the control strategies in self-excited induction generator is to regulate and maintain the required voltage and frequency of the generated output. This control is achieved using various types of controllers involving power electronics based devices. These controllers are called electronic load controllers. Electronic load controller in pico-hydro power plant using SEIG in isolated mode maintains the output load constant as seen by the SEIG under different load conditions. The SEIG

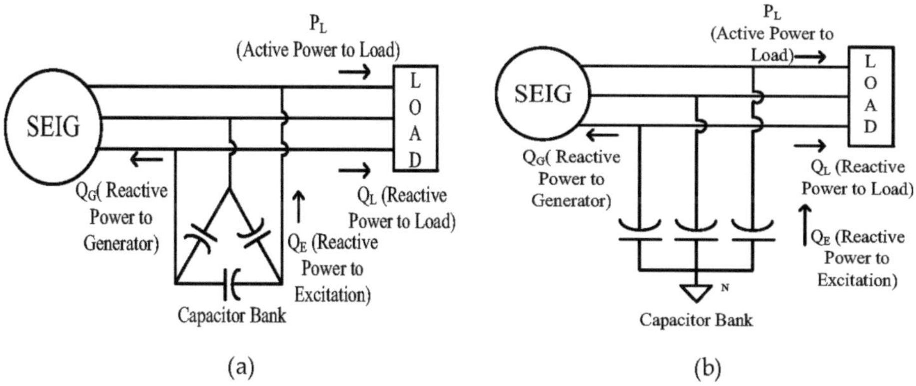

Figure 4.
Three phase SEIG with (a) delta connected capacitor bank, (b) star connected capacitor bank.

supplies output to two loads connected in parallel. The main load is consumer load and the other load is called dump load or auxiliary load. In this dump load, power dissipated is generally wasted. However, this can be used for other useful applications such as pumping of water etc. [3]. The chopper in this ELC circuit regulates the real & reactive power so that load seen by the generator is always constant. The conventional controller has discontinuous nature of control having discrete steps in connecting and disconnecting dump load. This results in power quality problems. The advanced electronic load controllers are designed to meet the demand of various types of consumer loads which may be of linear, non-linear, balanced or unbalanced type and also maintain the power quality in the system [11–19].

Conventional diode rectifier based ELC circuit consists of an uncontrolled 3-phase rectifier using diodes, a filtering capacitor, an IGBT or MOSFET based chopper circuit and resistive dump load as shown in **Figure 5**. The uncontrolled rectifier converts generated AC voltage at terminals of SEIG into DC voltage and filtering capacitor removes ac ripples from rectified output. The chopper circuit consists of an insulated gate bipolar transistor (IGBT) or MOSFET; which is used as electronic switch operated by a close loop controller based driver circuit. The controller circuit is used to switch ON or switch OFF the IGBT to connect the auxiliary dump load in such a way that total load seen by the SEIG terminal remains constant. However, the conventional ELC suffers from power quality problems. The improved or modern electronic load controllers have the capability of harmonics elimination, load balancing and maintaining desired voltage and frequency at various operating conditions. These improved electronic load controllers consist of voltage source converters (VSC) for voltage regulation, chopper circuit and auxiliary dump load. There are various topologies of these electronic load controllers based on the type of systems such as 3-phase-3 wire or 3-phase-4 wire systems to which SEIG is connected and also on the number of switches used in the VSC. One of the improved electronic load controllers which are the combination of conventional ELC and VSC based ELC

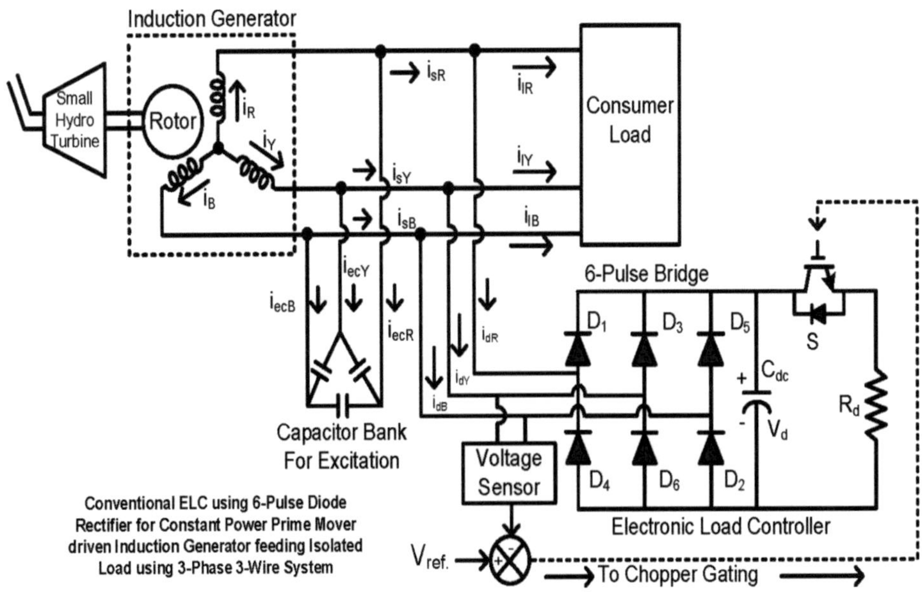

Figure 5.
Conventional ELC for isolated 3-phase SEIG in pico-hydro power generation system.

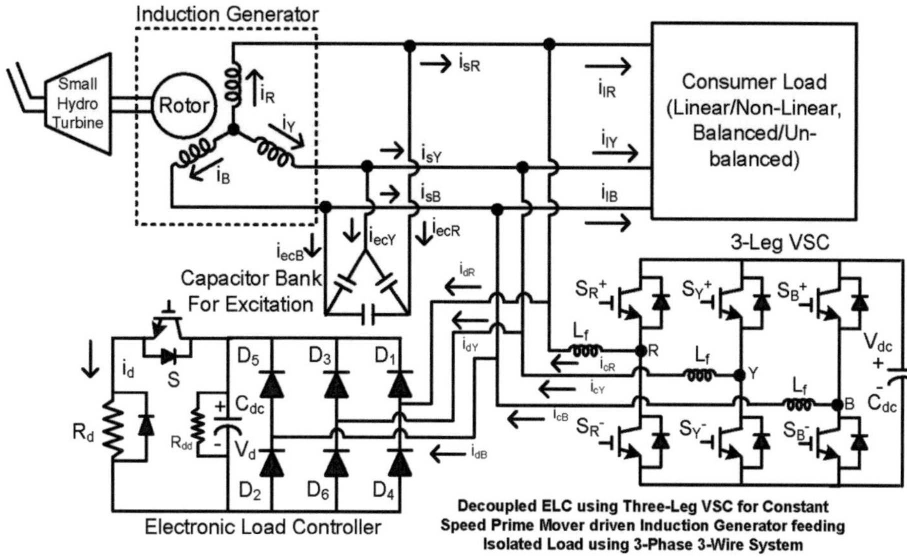

Figure 6.
De-coupled electronic load controller for 3-phase SEIG system feeding isolated load.

called de-coupled electronic load controller is shown in **Figure 6**. In this topology, ELC maintain the desired voltage & frequency like conventional controller, while the VSC serves the purpose of voltage regulation, eliminating harmonics and load balancing [16, 19].

3. Simulated performance evaluation of SEIG using MATLAB/Simulink

Before experimental validation of SEIG, performance characteristics of the SEIG under steady state condition were simulated using MATLAB-Simulink based model as shown in **Figure** 7 of a 3-phase induction machine of 2.2 kW, 415 Volts, 4-Poles rating which is actually used in real time practical application [20–23].

After selecting the required capacitance value in SEIG system under consideration for simulation, the capacitors are connected in delta mode in the model and various simulation tests are conducted under steady state conditions of rated voltage and speed. The different simulation conditions are created and the effect on capacitance rating is observed with variation in speed, variation in power factor, change in inductive load and various other loading conditions. Simulation results show the voltage build-up at a required capacitance and various of capacitance with speed. Results show that capacitance value increases with the decrease in power factor and also affected by the variation in connected load. For rated voltage, the value of excitation capacitance is slightly less than the value required at certain value of connected load. The terminal voltage of SEIG is affected by the change in load. As the load on SEIG increases, its terminal voltage decreases slightly. The MATLAB/Simulink based model and some of the results obtained from simulation such as developed terminal voltage and its variation at change in load, variation of capacitance at different load and power factors are shown in **Figures 8–11**. In the simulated model, the consumer load, controller circuit and measurement sections are shown as sub-systems [23].

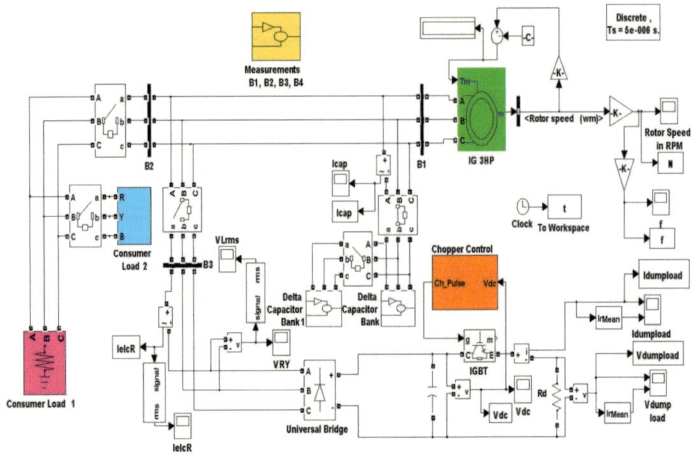

Figure 7.
MATLAB/Simulink based model of 3 HP, 3-phase, 415 V, 50 Hz SEIG in isolated pico-hydro power generation system.

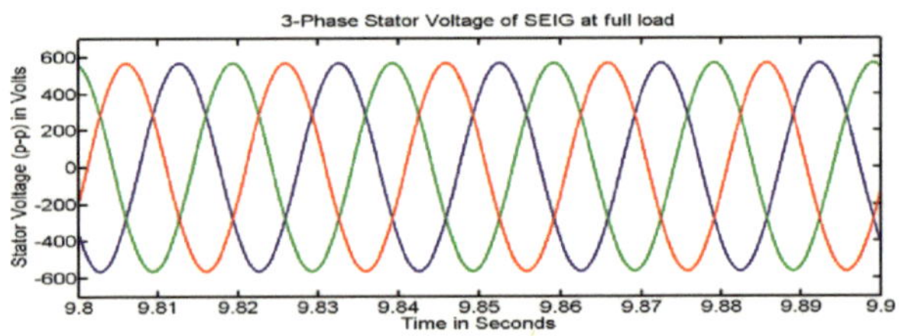

Figure 8.
Generated 3-phase output voltage of SEIG.

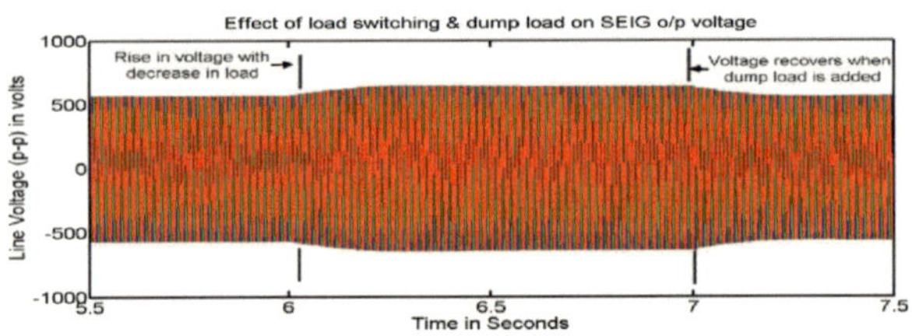

Figure 9.
Variation in terminal voltage with change in load.

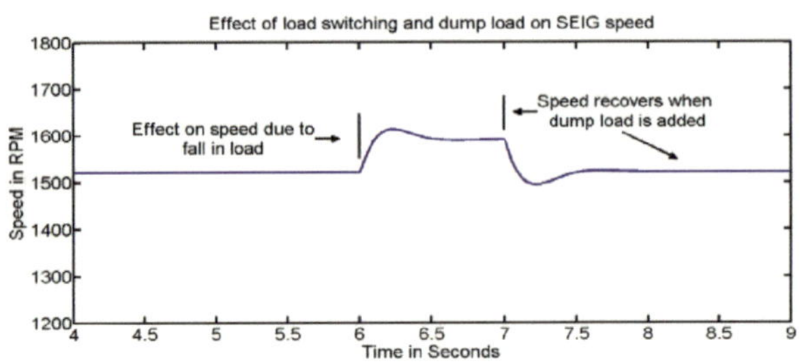

Figure 10.
Variation in terminal SEIG speed with change in load.

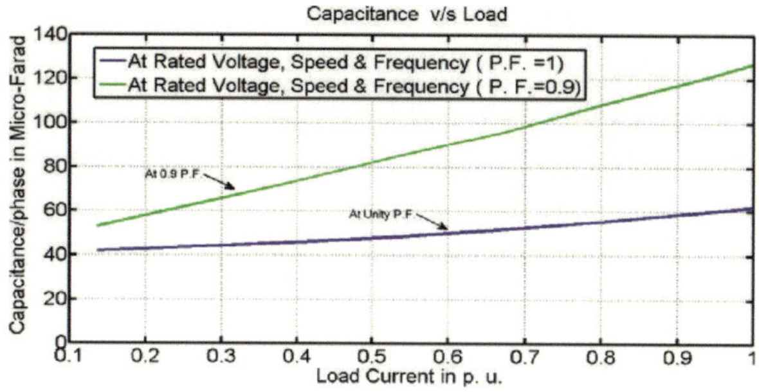

Figure 11.
Capacitance v/s load current curves.

4. Experimental evaluation of performance of 3-φ SEIG in pico-hydro power generation system

Simulated results are validated to prove the efficacy and suitability of self-excited induction generator feeding isolated load in pico-hydro power generation system using a hardware setup. **Figure 12** shows the complete experimental set-up in the lab to perform the experiments. The main components of hardware set-up are [23]:

- 3HP, 3-phase, 415 V, 4 pole, 50 Hz squirrel cage induction motor operated as 3-phase SEIG
- 5HP, DC shunt motor acting as prime-mover and coupled with induction machine
- Capacitor banks (3-phase/1-phase)
- Control Panels for induction machine and prime mover.
- Data logging, metering equipment and power meter for measurement of voltage, current, power, frequency, power factor and total harmonics distortion (THD)
- 3-phase resistive load panel containing lamp loads on each phase.
- Inductive load of 5A, 415 V rating.
- Electronic load controller circuit.
- Resistive dump load of 2.2 kW rating

The complete experimentation for operating the 3-phase induction motor as induction generator involves running the induction machine first as induction motor for few hours to initiate the residual magnetic field. Once it is sure that residual magnetism exists in the induction machine, it is the coupled with prime mover. To emulate the pico-hydro power plant which has constant power prime-mover characteristics, the selected machine as prime-mover should supply the constant power to the induction machine. In this set up, a DC shunt motor of 5 HP rating is selected to drive 3HP (2.2 kW) induction machine and its input power is monitored under all

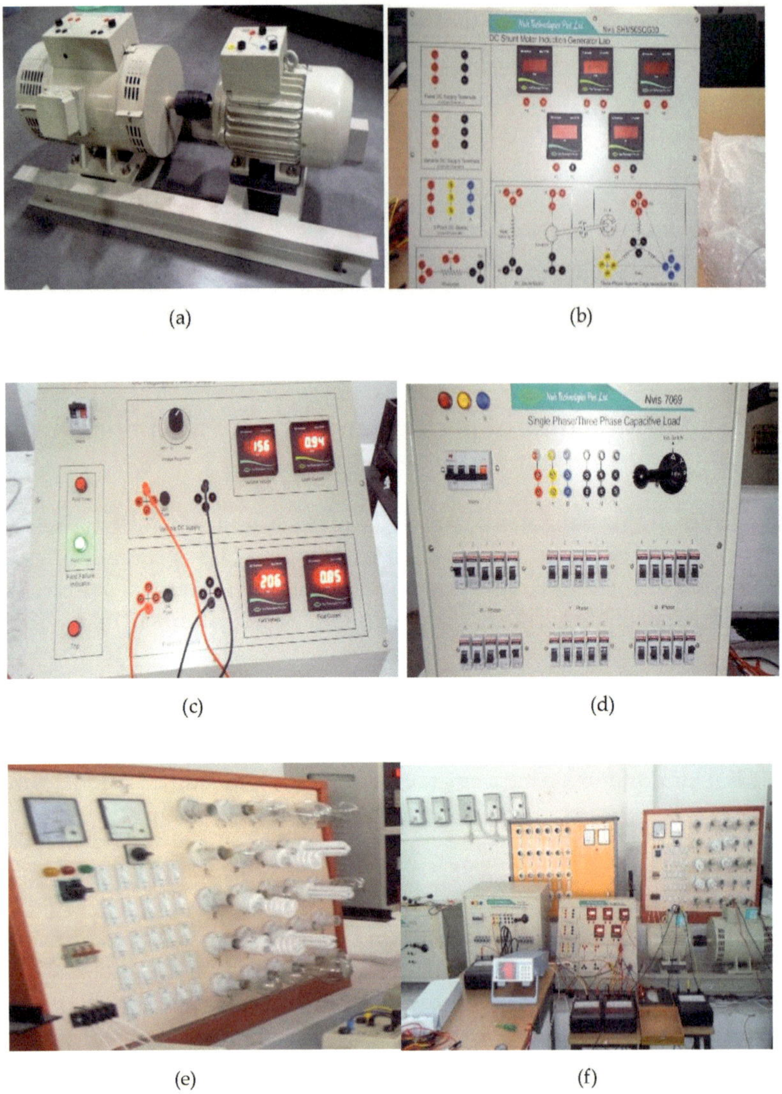

Figure 12.
Hardware setup: (a) induction machine coupled with prime-mover, (b) terminals and metering panel, (c) prime-mover motor control panel, (d) excitation capacitor bank panel, (e) load control panel (f) complete experimental set-up.

operating conditions. Now power to prime mover machine is switched on to drive the coupled induction machine to its rated speed. Prime-mover speed is further increased to take the 3-phase induction motor speed at above the synchronous speed (1500 RPM for 4-pole machine). Induction machine is driven at 1545 RPM for certain period without connecting the capacitor bank across SEIG terminals. The measurement devices are connected in the system at appropriate places to record the SEIG parameters at various excitation capacitance values and at different speeds. Initially the self-excited induction generator is run at no load. Then the first set of excitation capacitor bank (in delta or star mode) is switched ON to generate the no-load voltage across SEIG terminals. Then the value of the connecting capacitor banks is increased in steps until the required voltage is build up in the SEIG. And when the required

DOI: http://dx.doi.org/10.5772/intechopen.105650

3-φ, star connected resistive lamp load (W) across SEIG				Capacitance/phase in delta mode (μF)	Cap. bank line voltage (V)	Cap. bank line current (A)	Cap. bank phase current (A)	Calculated value of capacitance/phase (μF)	Frequency of generated voltage	SEIG speed in RPM
R_{ph}	Y_{ph}	B_{ph}	Total							
0	0	0	0	12	455	3.10	1.79	12.09	49.8	1525
100	100	100	300	12	430	2.89	1.67	11.9	49.8	1518
200	200	200	600	12	415	2.76	1.59	12.1	49.6	1510

Table 1.
Experimental values of excitation capacitance per phase in delta mode.

3-φ, star connected resistive lamp load (W) across SEIG				C/ph (μF) delta mode	V_{LL} (V)	$I_{L(C)}$ cap. line current (A)	I_{ph} (load) (A)	N (RPM)	PF	F (Hz)	Prime mover (DC shunt motor) parameters			
R_{ph}	Y_{ph}	B_{ph}	Total								V_A (V)	V_F (V)	I_A (A)	I_F (A)
0	0	0	0	6	0	0	0	1550	1	—	210	186	0.9	0.73
0	0	0	0	12	470	3.21	0	1552	1	50.8	209	185	2.5	0.70
0	0	0	0	12	415	2.66	0	1460	0.5	48.3	211	190	2.5	0.71
100	100	100	300	12	440	3.0	0.49	1539	0.90	50.3	208	184	4.3	0.69
200	200	200	600	12	425	2.82	0.95	1535	0.95	50.1	208	182	5.4	0.69
300	300	300	900	12	411	2.65	1.31	1530	0.94	49.8	207	183	6.9	0.68
340	0	320	660	12	445	3.0	1.35	1564	0.92	51.3	214	189	6.0	0.69
400	400	400	1200	12	383	2.50	1.52	1525	0.95	49.5	208	182	7.5	0.67
400	400	400	1200	12	417	2.65	1.60	1574	0.84	51.5	217	187	8.0	0.69
500	500	500	1500	12	355	2.30	1.77	1559	0.85	51	215	188	7.7	0.65
500	500	500	1500	18	415	2.7	1.9	1448	0.85	46.8	203	194	11	0.67

Table 2.
Experimental values of SEIG & prime-mover parameters under different operating conditions.

3-φ, star connected resistive lamp load (W) across SEIG				Parameters of 3-φ, star connected inductive load across SEIG			C/ph (μF) in delta mode	V_{LL} (V) SEIG terminal voltage	I_L (A) SEIG line current	$I_{L(C)}$ (A) cap. circuit line current	I_{ph} (load) (A) load phase current	N (RPM)	F (Hz)
Rph	Y_{ph}	B_{ph}	Total	V_{LL} (V)	I_L (A)	Q (kVAr)							
100	100	100	300	0	0	0	12	443	3.11	3.0	0.49	1540	50.2
100	100	100	300	452	1.20	0.95	18	452	3.39	3.02	0.49	1522	49.8
200	200	200	600	438	1.15	0.89	18	436	3.35	3.1	0.98	1519	49.7
400	400	400	1200	395	1.03	0.70	18	390	3.28	2.55	1.52	1512	48.8
400	400	400	1200	427	1.11	0.83	18	427	3.4	2.90	1.50	1542	50.2

Table 3.
Experimental values of SEIG parameters under different resistive & inductive loads.

voltage is build-up, load is connected across the SEIG and its effect on the SEIG speed, voltage, power factor and excitation capacitance is recorded under different loading and operating conditions.

4.1 Voltage build-up process in SEIG

At speed slightly higher than the synchronous speed of induction machine, a set of excitation capacitance (6 μF/phase) is switched ON and then the generated voltage at the terminals of SEIG is observed. In this set-up, full voltage development across the SEIG terminals is recorded when 12 μF/phase excitation capacitance in delta mode is connected for 2.2 kW rated induction generator at no load. Voltage generated at no load is slightly higher than the rated voltage of SEIG, but when load is switched, the voltage falls and is brought to its rated value by varying the excitation capacitance value connected across SEIG terminals. The required value of the capacitance as shown in **Table 1** can also be calculated by observing the terminal voltage and capacitor line current readings recorded during the experimentation. In this experimental process, the connected capacitance value in the capacitor bank connected in delta mode matched with the calculated value of excitation capacitance experimentally.

4.2 SEIG under loaded conditions

Domestic consumer electrical load in remote mountainous regions generally comprises of lighting and heating load using incandescent and compact fluorescent lamps. Apart from this; for a very short period electronic gadgets and house-hold single phase appliances are also used. To emulate these loading conditions, a three phase lamp load with multiple loading steps for each phase has been used to connect the lamp load on each phase in the experimental set-up. With greater emphasis on the maximum use of compact fluorescent lamps (CFLs) from energy conservation point of view, CFLs have also been used along with conventional incandescent lamps to evaluate the performance of SEIG. **Table 2** shows the experimental recorded parameters [23].

Experimental results show that a minimum value of excitation capacitance is required to build up the voltage at SEIG terminals on no load. For 2.2 kW, 4-pole SEIG, the terminal voltage built-up at 12 μF/phase in delta connected capacitor bank and it will be 36 μF/phase if connected in star connected mode [23]. With the increase in load, the SEIG terminal voltage slightly decreases along with the slight reduction in SEIG speed. This decrease in terminal voltage of SEIG due to increased load can however be balanced by increasing prime mover speed. Further increase in electrical load leads to increased requirement of excitation capacitance at rated speed. Since most of the domestic consumer electrical load is of single-phase type, 4-wire electrical distribution system is used for electrifying the villages which results in the un-balancing of load in each phase. To emulate this, unequal loading of SEIG has been done in the experimental study. The requirement of the excitation capacitance increases when inductive load is added to compensate for the lagging reactive power caused by the inductive load. **Table 3** shows the SEIG parameters recorded under resistive and inductive load.

5. Conclusions

The simulated performance of 3-phase SEIG system feeding isolated load have been investigated in this chapter and results are further validated using an experimental setup. The observed results validate the usefulness and suitability of self-excited induction generators for feeding the isolated load in remote mountainous regions as it adapts well to meet the varying loading requirements and keeping the voltage and frequency of the system under permissible limits. During less load period, the wasted energy across dump load can further be utilized for other useful applications such as pumping of water to meet the potable and irrigation requirement in the hilly terrains. The vast use of these self-excited induction generator based systems can be useful to harness the maximum small hydro potential without affecting the environment much and also to meet the electricity requirement of remote Himalayan mountainous regions where distributing electricity is itself a challenging task due to harsh climatic and tough terrain conditions.

Acknowledgements

Our sincere thanks to the management of Himachal Pradesh State Electricity Board Limited to collect the daily electrical load data from their 33 kV/11 kV substation feeders feeding the remote locations presented in this paper.

Author details

Umesh C. Rathore[1*] and Sanjeev Singh[2]

1 Electrical Engineering Department, Government Hydro Engineering College Bandla, Bilaspur, Himachal Pradesh, India

2 Electrical Engineering Department, MA NIT Bhopal, Madhya Pradesh, India

*Address all correspondence to: rathore7umesh@gmail.com

References

[1] Simoes MG, Farret FA. Alternate Energy Systems: Design and Analysis with Induction Generators. 2nd ed. Boca Raton: CRC Press; 2008

[2] Rathore UC, Singh S. Performance evaluation of 3-phase self-excited induction generator for remote mountainous region of Himalayas. In: Proceedings of the IEEE International Conference on Control, Instrumentation, Energy & Communication (CIEC). India: University of Calcutta; 2014. pp. 421-425

[3] Rathore UC, Singh S. Isolated 3-phase self-excited induction generator in pico-hydro power plant using water pump load in mountainous region of Himalayas. In: Proceedings of 2014 IEEE Global Humanitarian Technology Conference - South Asia Satellite (GHTC-SAS). 2014. pp. 40-44

[4] Murthy SS, Malik OP, Tandon AK. Analysis of self-excited induction generator. In: Proceedings of the. Inst. Elect. Eng. C. 1982. pp. 260-265

[5] Singh SP, Singh B, Jain MP. Comparative study on the performance of a commercially designed induction generator with induction motors operating as self-excited induction generators. Proceedings of IEE–C. 1993;**140**(5):374-380

[6] Bansal RC. Three-phase self-excited induction generators: An overview. IEEE Transactions on Energy Conversion. 2005;**20**:292-299

[7] Al Jabri AK, Alolah AI. Capacitance requirement for isolated self-excited induction generator. Proceedings of IEE Electric Power Application. 1990;**137**(3):154-159

[8] Chan TF. Capacitance requirements of self-excited induction generators. IEEE Transactions on Energy Conversion. 1993;**8**(2):304-311

[9] Shridhar L, Singh B, Jha CS, Singh BP, Murthy SS. Selection of capacitors for the self-regulated short shunt self-excited induction generator. IEEE Transactions on Energy Conversion. 1995;**10**(1):10-17

[10] Chandran VP, Vadhera S. Capacitance requirements of self-excited induction generator for different operating conditions. In: Proceedings of the IEEE ICEAS. 2011. pp. 1-6

[11] Suarez E, Bortolotto G. Voltage-frequency control of a self-excited induction generator. IEEE Transactions on Energy Conversion. 1999;**14**(3):394-401

[12] Singh B, Murthy SS, Gupta S. Analysis and design of STATCOM-based voltage regulator for self-excited induction generators. IEEE Transactions on Energy Conversion. 2004;**19**(4):783-790

[13] Rathore UC. Design and Development of Voltage and Frequency Controller for Micro/Pico Hydro Power Generation System. Punjab; India: SLIET University Longowal; 2017

[14] Singh B, Murthy SS, Gupta S. Transient analysis of self-excited induction generator with electronic load controller (ELC) supplying static and dynamic loads. IEEE Transactions on Industry Applications. 2005;**41**(5):1194-1204

[15] Singh B, Kasal GK. Voltage and frequency controller for isolated asynchronous generators feeding 3-phase

4-wire loads. In: Proceedings of the IEEE ICIT'06. 2006. pp. 2773-2778

[16] Singh B, Kasal GK. Analysis and design of voltage and frequency controllers for isolated asynchronous generators in constant power applications. In: Proceedings of the IEEE PEDES'06. 2006. pp. 1-7

[17] Ramirej JM, Torres E. An electronic load controller for self-excited induction generator. Proceedings of the IEEE PES. 2007;**24-28**:1-8

[18] Idjdarene K, Rekioua D, Rekioua T, Tounzi A. Performance of an isolated induction generator under unbalanced loads. IEEE Transactions on Energy Conversion. 2010;**25**:303-311

[19] Rajagopal V, Singh B. Improved electronic load controller for off grid induction generator in small hydro power generation. In: Proceedings of the IEEE IICPE'10. 2011. pp. 1-7

[20] Murthy SS, Ahuja RK. Design and analysis of three phase self-excited induction generator using MATLAB GUI based methodology. In: Proc. IEEE ICPECS'10. 2010. pp. 357-361

[21] Cozorici F, Vadan I, Munteanu R A, Cozorici I, Karaissas P. Design and simulation of a small wind-hydro power plant. In: Proceedings of the IEEE ICCEP'11, Ischia. 2011. pp. 308-311

[22] Scherer LG, Camargo RF, Pinheiro H, Rech C. Advances in modelling and control of micro hydro power stations with induction generators. In: Proceedings of the IEEE ICCEP'11. 2011. pp. 997-1004

[23] Rathore UC, Singh S. Experimental evaluation of performance of constant power prime-mover driven isolated 3-phase SEIG for pico-hydro power generation system in remote mountainous region of Himalayas. In: Proceedings of the ICAET-2016, MATEC Web of Conferences, India. pp. 1-6

Section 3

Renewable Energy Materials

Chapter 5

Engine Lightweighting: Use of Green Materials as Reinforcement in Aluminum Metal Matrix Composites

Akaehomen O. Akii Ibhadode

Abstract

Lightweighting of automobiles of which the IC engine is a part has become very important due to stringent emission regulations being imposed on vehicle manufacturers, and the need to have more fuel-efficient vehicles. The use of light weight materials such as aluminum metal matrix composites (AMMCs) made up of aluminum alloy and nonmetal reinforcements such as alumina and silicon carbide is one strategy used for lightweighting. Recently, there has been active research in the use of biodegradable green materials such as agricultural wastes as reinforcements for AMMCs. In this chapter, work done on the use of biodegradable green materials as reinforcements for AMMCs is reviewed. The potential for their use as engine parts materials is analyzed. The results show that they have the potential to provide significant weight and cost savings when used as engine parts materials.

Keywords: lightweighting, aluminum metal matrix composites, IC engine, green reinforcement materials, biodegradable green materials

1. Introduction

Automobile lightweighting involves reducing the weight of an automobile in order to minimize fuel consumption and exhaust emissions. In electric vehicles (EV) however, the effect of lightweighting is to extend the range of each battery charge. The Vehicle Technologies Office of the United States Office of Energy Efficiency and Renewable Energy [1] states that “a 10% reduction in vehicle weight can result in a 6%-8% fuel economy improvement”. Also, *Järvikivi* [2] reported in 2021 that each 100 kg reduction in a vehicle’s weight cuts (on average) 8.5 g of CO_2 per 100 km. These figures tell us that a vehicle’s performance with respect to fuel efficiency and cleaner operation is improved by lightweighting of the vehicle. The engine, which is the powerhouse of the vehicle can, on the average, weigh around 10% of vehicle curb weight. Thus, anything done to reduce the engine weight will impact the vehicle performance positively.

Strategies for lightweighting engines include the use of high specific strength materials (that is, high strength-to-weight ratios) such as high-strength aluminum alloys or aluminum metal composites, sheet metal fabrication of parts, innovative design of parts, additive manufacturing and friction reduction of moving parts [3–7].

Metal composites are engineered materials consisting of a metal matrix in which a nonmetal reinforcement is dispersed. The resulting material called a metal composite, combines the properties of the metal and the reinforcement material to produce better desirable properties than any of the two can provide individually. In this way, low-strength, lightweight engineering materials such as aluminum are converted to relatively high-strength materials suitable for use in fabrication of engine parts and others in the automobile. Common reinforcement materials include silicon carbide (SiC), aluminum oxide (Al_2O_3), titanium dioxide (TiO_2), etc. [8–10].

With the advent of the Circular Economy [11, 12] and the necessity of going Green [13, 14], there has been a recent move to use green materials, especially biodegradables, essentially agricultural wastes, as reinforcement materials in metal matrix composites [15, 16]. Green materials are materials that have the following characteristics: are local and renewable, nontoxic, improve occupancy health, lowers cost, and conserve energy and water use and waste products, and have low embedded energy in their harvesting or collection, production, transportation, and use [17]. Examples of biodegradable green materials include palm kernel shells, periwinkle shells, coconut shells, corn cob, bagasse, rice husk, egg shell, etc. This has a number of potential benefits: more sustainable production, lower cost of reinforcement materials, optimal use of resources including waste materials, and cleaner environment in places where these biodegradable wastes are produced.

This chapter reviews the work done in using biodegradable green materials as reinforcements in aluminum metal matrix composites (AMMCs). It discusses their application to IC engine parts manufacture.

2. Review of green materials reinforced aluminum metal matrix composites (AMMCs)

In recent times there has been much interest in using renewable materials in the form of waste agricultural products as reinforcement materials in AMMCs, especially in developing countries. This is a result of a number of reasons: (i) they serve as a cheap alternative to conventional reinforcing materials such as silicon carbide and alumina, which are imported into these countries, and are expensive, (ii) enables the possibility of producing locally in these countries, lightweight high strength composites, and (iii) has the potential to create a cleaner environment due to the usage of these waste agricultural products for useful products.

The Appendix **Table A1** gives a summary of some of the research works carried out in using green materials for reinforcing AMMCs [18–85].

3. Discussion

3.1 Range of green reinforcement materials

Table 1 lists the green materials used by various researchers as reinforcement for AMMCs. It shows that green reinforcement materials cover a wide range of materials, mainly agricultural wastes. Twenty green reinforcement materials are shown.

Bagasse	Bamboo leaf	Bone	Breadfruit seed hull	Coconut shell
Corncob	Cow horn	Egg shell	Groundnut shell	Lemon grass
Maize stalk	Mango seed	Marula seed shell	Neem leaf	Palm kernel shell
Palm sprout	Periwinkle shell	Rice husk	Snail shell	Tamarind leaf

Table 1.
Green materials for reinforcing AMMCs as used by researchers.

3.2 Preparation of green reinforcement materials

Figure 1 shows the processing routes of biodegradable green materials for reinforcement of AMMCs derived from the review summary in the Appendix **Table A1**. The preparation starts with cleaning with water or distilled water to remove all impurities. This is followed by drying, usually in the sun for days or dried in an oven for a few hours. The dried material may be pulverized and then ashed; or ashed directly

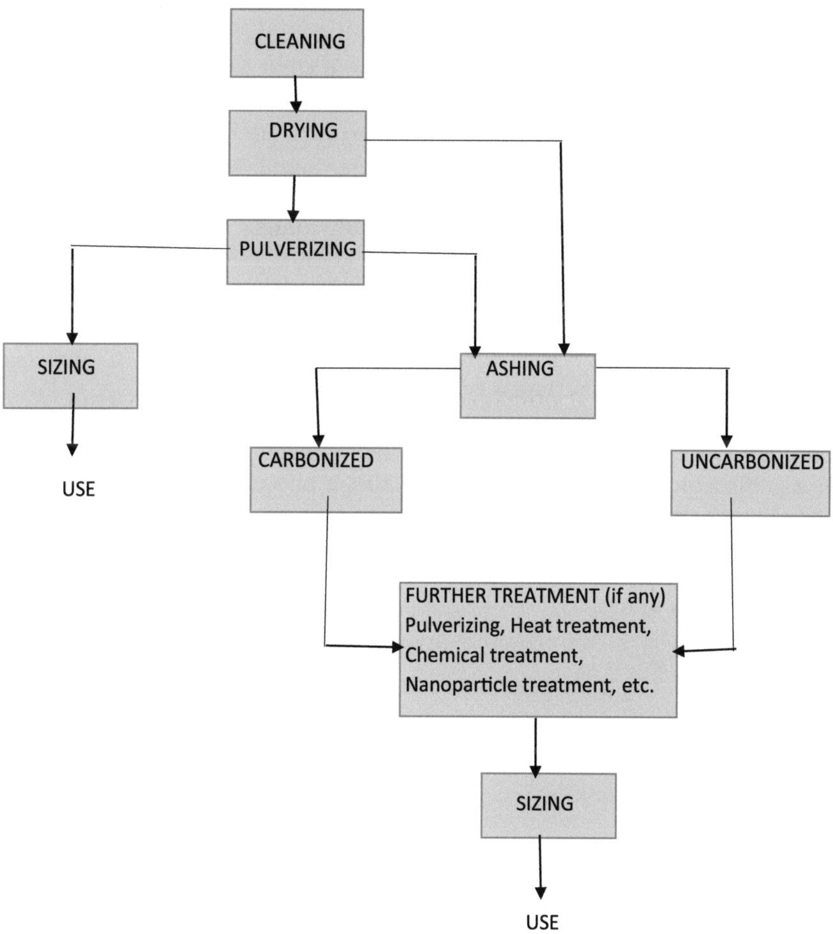

Figure 1.
Preparation routes for green reinforcement materials.

without pulverization. Depending on what is desired, the pulverized particles after sizing are used directly for reinforcement. Ashing can be done by heating in air or without air. When heated without air, carbonized ash is obtained. When heated with air, uncarbonized ash is obtained. The obtained ash may further be treated before sizing for use as a reinforcement material. Further treatment could include, pulverizing, conditioning by heat treatment in a furnace to reduce the carbonaceous and volatile constituents [77], chemical treatment and/or reduction to nanoparticle treatment [63]. Finally, the particles are sized by sieving and thereafter used as reinforcements.

The use of pulverized particles without ashing is a simpler route and most likely, less costly. There seems not to be any detailed work on the comparison of the performance of metal composites reinforced with straight particles, uncarbonized and carbonized ashes. However, Hassan and Aigbodion [44] showed that carbonized eggshell ash had better performance than uncarbonized ash.

3.3 Chemical compositions of green reinforcement materials

Table 2 shows the chemical compositions of most of the green reinforcement materials listed in **Table 1**. We could classify the reinforcements whose oxide analyses are given into (i) silica-based, (ii) calcium oxide-based, and (iii) alumina/ferric oxide-based reinforcements. These are shown in **Table 3**. Eight of the materials are silica based having at least 41% SiO_2 as their highest oxide contents, three are calcium oxide based having at least 40% CaO as their highest oxide contents, and only one is alumina/ferric oxide based having at least 35% Al_2O_3 and at least 30% Fe_2O_3. When these green materials are used as reinforcements in AMMCs, they release their oxides into the matrix to affect the properties of the metal matrix.

Table 2 also lists the densities of these biodegradable green materials, which shows that most of the materials have much lower densities than aluminum which is commonly used as matrix of metal composites.

3.4 Fabrication of green materials reinforced AMMCs

There are a number of methods used for the fabrication of AMMCs [8]. However, the stir casting process is frequently used because it promotes the casting of uniformly reinforced metal composites as stirring transfers particles into the liquid metal and maintains the particles in a state of suspension [33]. The two-step stir-casting process involves the following steps [99]:

i. Preheating of the reinforcement particles to improve wettability with the metal matrix

ii. Heating the metal matrix above the liquidus temperature

iii. Cooling the liquid metal matrix in the furnace to a semi-solid state

iv. Adding the preheated reinforcement particles to the semi-solid metal matrix and stirring continuously for a short period, say 5 min

v. Superheating the composite slurry while stirring at a specified speed and time

vi. Casting the molten composite into molds

Green material	SiO_2	Al_2O_3	Fe_2O_3	CaO	MgO	SO_3	K_2O	Na_2O	MnO	ZnO	TiO_2	Cr_2O_3 or (Cl)	P_2O_5	Mn_2O_3 Or (NiO)	SrO	LOI	Density, g/cm^3
Rice Husk Ash [22]	97.095	1.135	0.316	0.073	0.825	0.146	0.181	0.092	—	—						0.965	1.8–2.1 [86]
Bamboo Leaf Ash [25]	75.9	4.13	1.22	7.47	1.85		5.62				0.20						2.64 [87]
Bagasse Ash [22]	77.286	10.951	3.660	2.088	1.489	0.487	3.159	0.381	—	—						3.277	0.238 [22]
Breadfruit Seed Hull Ash [33]	15.45	35.80	30.34		1.20		0.52	0.45	0.22	0.05		5.06					1.98 [33]
Coconut Shell Ash [29]	46	16	14		18	.	1.2	0.9	0.5	0.6							2.05 [37]
Corncob Ash [88]	66.34	7.48	4.44	11.57	2.06	1.07	4.92	0.41									0.8–1.2 [89]
Egg Shell [90]	0.09	0.03	0.02	50.70	0.01	0.57		0.19				(0.219)	0.24	(0.001)	0.13	47.8	2.49 (uncarbonized) 1.98 (carbonized) [90]
Mango Seed Shell Ash [56]	43.574	13.847	6.211	14.895	3.173	1.797	4.559	2.302		2.653	2.129	0.005	4.173	0.130	2.653		0.4–0.425 [91]
Groundnut Shell Ash [50]	41.42	11.75	12.60	11.23	3.51	0.44	11.89	1.02	0.23	—	0.63	—	1.71			3.57	0.23 [92]
Palm Kernel Shell (unashed) particle [61]	55.69	9.43	3.32	11.21	4.85	0.67	9.71	1.76	0.72	—	—	0.27	2.39				1.70–2.05 [93]
Periwinkle Shell [31]	32.84	10.20	7.02	40.84	1.47	0.26	0.14	0.24			1.07		(P_2O = 0.01)	0.78			2.3 [70]
Snail Shell Ash [94]	0.55	1.35	0.4	74.30	0.62		0.21	0.25		0.01	0.01	0.001	0.22				2.55–2.81 [95]
	Ca	Mg	K	P	CO	Organic carbon	S	N									
Cow bone [31]	36.05	0.74	0.85	16.43	4.58												1.24–1.71 [96]
Cow hone [97]	0.396	0.014	0.09	0.165		39.97	0.049	15.54									1.283 [98]

Table 2.
Chemical compositions of green reinforcement materials.

	Rice husk	Bagasse	Bamboo leaf	Corncob	Palm kernel shell	Coconut shell	Mango seed shell	Groundnut shell	Snail shell	Egg shell	Periwinkle shell	Breadfruit seed shell
Silica based	>97% SiO_2	>77% SiO_2	>75% SiO_2	>66% SiO_2	>55% SiO_2	>46% SiO_2	>43% SiO_2	>41% SiO_2				
Calcium oxide based									>74% CaO	>50% CaO	>40% CaO	
Alumina/Ferric oxide based												>35% Al_2O_3/ >30% Fe_2O_3

Table 3.
Classification of green reinforcement materials according to their prominent oxide content.

3.5 Microstructure of green materials reinforced AMMCs

The large body of knowledge on metal matrix composites as exemplified by the references on green materials reinforced AMMCs in the Appendix **Table A1** shows that when the metal composite is well fabricated, the microstructure of the composites shows the uniform distribution of green materials particles in the aluminum matrix. The uniform distribution of the reinforcement particles in the microstructure of the composites is the main reason for the enhancement of the mechanical properties of AMMCs [33, 44, 100]. Also, Atuanya et al. [33] showed that the increase in reinforcement weight fraction gives rise to decrease in matrix grain size in the composites as shown in **Figure 2**. This decrease in matrix grain size further improves the mechanical properties of the composites as reinforcement content increases.

3.6 Physical and mechanical properties of green materials reinforced AMMCs

Figure 3 shows the plots of relative density ρ_r, relative tensile strength σ_r, relative hardness H_r, and relative impact energy E_r of aluminum metal matrix composites (AMMCs) against weight percent of reinforcement particles, for egg shell ash [44] and breadfruit seed hull ash [33].

The relative density, relative tensile strength, relative hardness, and relative impact energy are defined respectively as

$$\rho_r = \rho_c/\rho_m \tag{1}$$

where ρ_m = density of reference metal matrix, ρ_c = density of composite.

$$\sigma_r = \sigma_c/\sigma_m \tag{2}$$

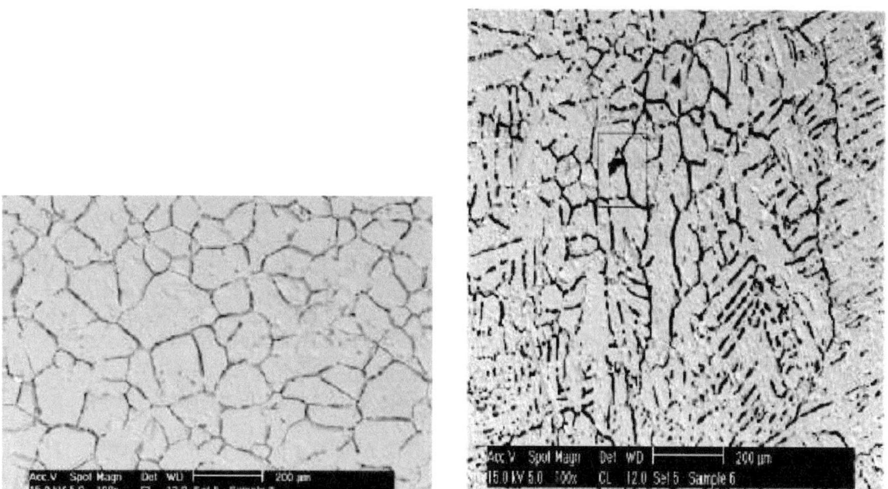

(a) (b)

Figure 2.
SEM of aluminum alloy reinforced with (a) 2 wt.% and (b) 10 wt.% of breadfruit seed hull ash (mag. ×100) [33].

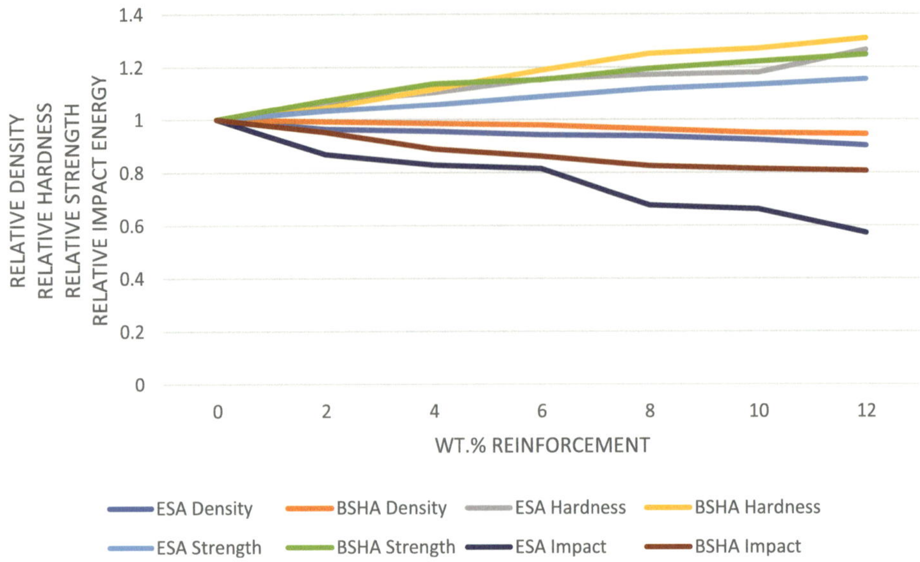

Figure 3.
Density, hardness, strength and impact energy of aluminum metal matrix composites reinforced with egg shell ash and breadfruit seed hull ash.

where σ_m = tensile strength of reference metal matrix, σ_c = tensile strength of composite.

$$H_r = H_c/H_m \tag{3}$$

where H_m = hardness of reference metal matrix, H_c = hardness of composite.

$$E_r = E_c/E_m \tag{4}$$

where E_m = impact energy of reference metal matrix, E_c = impact strength of composite.

The figure shows that as the weight percent of reinforcement materials increases, density and impact strength decrease, while tensile strength and hardness increase for both reinforcement materials.

The density of composites decreased as the percent addition of reinforcement materials increased because the green materials, egg shell ash (ρ = 1.98 g/cm^3), breadfruit seed hull ash ((ρ = 1.98 g/cm^3) (see **Table 2**) are light materials compared to the densities of the aluminum metal matrices used (2.775 g/cm^3 in both cases). This is in contrast to the use of conventional reinforcement materials such as alumina and silicon carbide. **Figure 4** shows that as the addition of alumina ($\rho_{alumina}$ = 3.95 g/cm^3) and silicon carbide (ρ_{SiC} = 3.21 g/cm^3) increased, the density of the composites increased because of their higher densities than the reference aluminum alloy matrix. This shows that use of biodegradable green materials as reinforcement materials for AMMCs will produce lighter composites suitable for lightweighting of IC engines and other automobile parts, and industrial parts where lightweighting is important.

Figure 3 shows that the hardness of the composites increased, above the reference metal matrix, as percent reinforcement content increased. This is attributed to an increase in the volume of precipitated phases or a high dislocation density. The

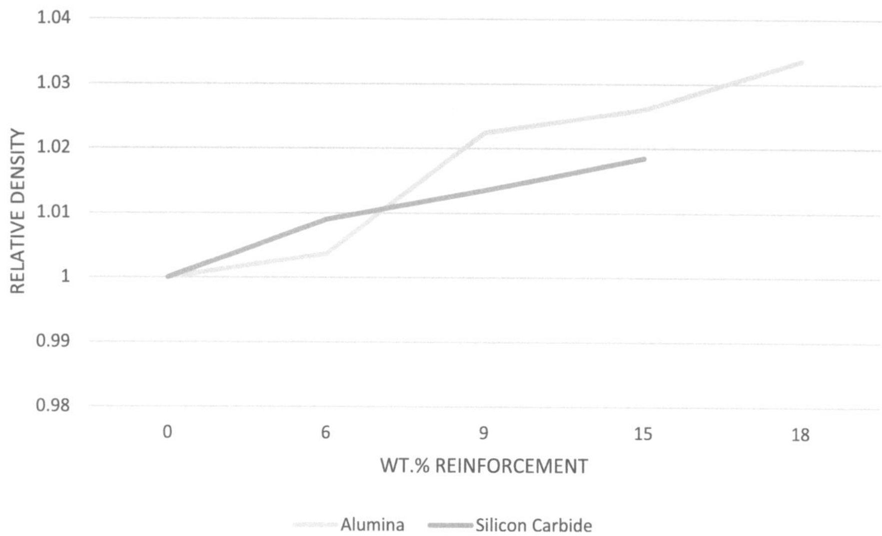

Figure 4.
Relative densities of AMMCs reinforced with alumina [99] and silicon carbide [101].

increase in the weight percentage of hard and brittle phases of the reinforcing materials in the aluminum alloy is responsible for the increments in hardness as reinforcement materials increase [33].

Also, **Figure 3** shows that as the percent weight of reinforcement increased, the tensile strength increases over the reference metal matrix. This may be a result of several factors such as good particle/matrix interfacial bonding, fine reinforcement particle size, and the strengthening effect of the reinforcing materials [33, 102–104].

The figure also shows that as the percent reinforcing material increased, the impact strength of the composites decreases. This is due to the brittle nature of the reinforcing materials which degrades the impact strength as the reinforcement material increases [33].

3.7 Application of green materials reinforcements to IC engine parts

Table 4 shows IC engine parts in which aluminum alloys are used. The table shows the permissible stresses specified for the parts. The strengths of some biodegradable green reinforcement materials are also shown to see if they measure up with strength requirements of the engine parts. These strength figures indicate that there is promise for use of these biodegradable green materials for reinforcement of AMMCs as IC engine parts materials. We note that their strength values are increased when high aluminum alloys are used as reference metal matrix. These results indicate that these biodegradable reinforcement materials have the latitude of being able to generate high-strength AMMCs by choice of type of aluminum alloy, type of biodegradable material, and amount of reinforcement. By this statement, it follows that all the IC engine parts shown in **Table 4**, and possibly, other parts not shown, could readily be made with biodegradable green materials reinforced AMMCs. We note that factor of safety values which could be as high as 6 in certain cases, such as in connecting rod design, are required to reduce the strength values of the green biodegradables shown in **Table 4**. Despite this, biodegradables show great promise as reinforcing materials in AMMCs.

Engine part	Permissible stress, MPa [105]	Tensile strength of biodegradable reinforced AMMC, MPa				
		Rice husk [74]	Palm kernel shell [63]	Periwinkle shell [70]	Breadfruit seed hull [33]	Egg shell [44]
Engine block	35–100	312.5	263.4	132.5	215	113
Cylinder head	30–50					
Piston	50–90					
Connecting rod	60–100					

Key: Rice Husk (A356.2/8wt.%RHA) [74], Palm Kernel Shell (A356/4wt.%PKSA) [63], Periwinkle Shell (Al6063/15wt.%PSA) [70], Breadfruit Seed Hull (Al–Si–Fe/12wt.%BSH) [33], Egg Shell (Al-Cu-Mg/12wt.%EG) [44].

Table 4.
Potential application of biodegradable materials reinforced AMMC to IC engine parts.

3.8 Engine lightweighting by biodegradable green materials reinforced AMMCs

From previous work [5], in which a gasoline engine rated at 7.1 kW at 5500 rpm generating a maximum torque of 18.0 Nm at 3500 rpm was used to calculate engine weight reduction, we use this same engine to calculate potential weight reductions when using is made of these biodegradable green materials reinforced AMMCs to replace the materials used to make some parts namely, engine block, cylinder head, piston, and connecting rod. **Table 5** shows the engine weight reduction analysis. The engine block was made of cast iron, and cylinder head, piston, and connecting rod made of aluminum alloys were replaced in the analysis with AMMCs reinforced with rice husk (RH) ash, palm kernel shell (PKS) nanoparticles, and periwinkle shell (PS) ash. The table shows that:

i. Total parts (engine block, cylinder head, piston, and connecting rod) weight reductions are 54.63%, 61.54%, and 63.01% when PS, RH, and PKS reinforced AMMCs are used respectively in descending order of reinforcement density.

ii. Total engine weight reductions are 28.64%, 32.26%, and 33.04% when PS, RH, and PKS-reinforced AMMCs are used respectively.

The table also shows when it is assumed that aluminum alloy A356 is used to originally make the engine block. With this scenario, the total parts weight reductions are 13.94%, 27.03%, and 29.83% when PS-, RH-, and PKS-reinforced AMMCs are used, respectively. Also, in this case, the total engine weight reductions are 5.12%, 9.93% and 10.96% when PS-, RH-, and PKS-reinforced AMMCs are used, respectively.

These results show the massive engine lightweighting achieved when only a single cast iron part is replaced with AMMC. When the engine block was assumed to have been made originally from aluminum alloy A356, the weight reductions achieved by replacement with the biodegradable green materials reinforced AMMCs, are also appreciable: the total weight reductions on parts are 13.94%, 27.03%, and 29.83%, and on whole engine are 5.12%, 9.93%, and 10.96% when reinforced with PS, RH, and PKS, respectively.

These results follow the trend of density reduction. The percent weight reduction depends on the percent reduction of densities from original engine material to the biodegradable green reinforcement material. For example, the density reduction from

Engine part	Material	Density, kg/m^3 [Ref?]	Weight, kg	Volume, m^3	Weight Equivalent of Biodegradable Reinforced AMMC, kg			Weight Reduction, kg/(% on individual parts)		
					RH, (ρ_c = 1.95 g/cm^3)	PS (ρ_c = 2.3 g/cm^3)	PKS, (ρ_c = 1.875 g/cm^3)	RH	PS	PKS
Engine Block	Cast Iron	7079	10.25	0.001448 (1448 cm^3)	2.824	3.330	2.715	7.426/ (72.45%)	6.92/ (67.51%)	7.535 (73.51%)
Cylinder Head	Aluminum Alloy (A356)	2670	3	0.001124 (1124 cm^3)	2.192	2.585	2.108	0.808/ (26.93%)	0.415/ (13.83%)	0.892/ (29.73%)
Piston	Aluminum Alloy (A4032)	2690	0.094	0.000035 (35 cm^3)	0.068	0.081	0.066	0.026/ (27.66%)	0.013/ (13.83%)	0.028/ (29.79%)
Connecting Rod	Aluminum Alloy (A7075)	2803	0.157	0.000056 (56 cm^3)	0.109	0.129	0.105	0.048/ (30.57%)	0.028/ (17.83%)	0.052/ (33.12%)
Total			13.501	0.002663 (2663 cm^3)	5.193	6.125	4.994	8.308/(−)	7.376/(−)	8.507/(−)
Weight reduction on total parts of 13.501 kg								61.54%	54.63%	63.01%
Weight reduction on total parts of 7.117 kg if engine block was originally made from A356 (details not shown)								27.03%	13.94%	29.83%
Weight reduction on total engine weight of 25.75 kg								32.26%	28.647%	33.04%
Weight reduction on total engine weight of 19.366 kg if engine block was originally made from A356 (details not shown)								9.93%	5.12%	10.96%

Table 5.
Engine weight reductions by biodegradable materials reinforced AMMCs.

aluminum alloy A356 to palm kernel shell (PKS) is about 30% which is the same percent reduction in weight. Thus, finding a low-density reinforcing material with high reinforcement performance is the key to weight reduction.

3.9 Cost analysis of biodegradable materials reinforced AMMCs production

A cost analysis of the production of one of the biodegradable reinforcement materials discussed above, specifically, PKS is presented. This is done to quantify the financial benefit that may accrue to the deployment of waste biodegradables as reinforcing materials in AMMCs.

Refer to **Figure 1** which shows the processing steps for obtaining reinforcement particles for AMMCs. For the analysis, we assume PKS particles, unashed and uncarbonized. Unashed and uncarbonized PKS particles have been used directly to reinforce AMMCs successfully by Edoziuno et al. [61], Ibhadode and Ebhojiaye [8].

In manufacturing, the material cost C_m is usually a fraction of the production cost, C_p. That is:

$$C_m/C_p = x_m < 1 \tag{5}$$

To simplify the analysis, we make the following assumptions:

i. we take a moderately low value of $x_m = 0.2$, due to uncertainties that may be associated with the manufacturing process such that four-fifths of the production cost is attributed to other input costs such as energy, equipment, overheads, and others. This assumption is justified because PKS is a waste product whose cost may not be greater than the cost of collection

ii. cost of PKS, C_m, is equal to the cost of collecting it as a waste product from palm kernel oil (PKO) processing. PKO mills are supplied with palm kernel seeds cracked by small businesses which discard the shells as waste. This collection cost depends on the distance between the points of collection and usage. If we assume a collection radius of 30 km, going by existing prices, cost of delivering a 10-ton truck load may be about N50,000. Thus, PKS cost per kg is shown as follows:

$$C_m = \text{N } 50{,}000/(10 \text{ tons} \times 1000 \text{ kg}) = \text{N}5/\text{kg} \tag{6}$$

From Eq. (5), for $x_m = 0.2$, production cost is

$$C_p = C_m/x_m = \text{N}5/0.2 = \text{N}25/\text{kg} \tag{7}$$

If we assume a profit margin of 25%, then the selling price of PKS particles is as follows:

$$C_s = \text{N}25/\text{kg}\,(1 + 0.25) = \text{N}31.25/\text{kg} \tag{8}$$

Thus, a ton of PKS reinforcement particles will cost N31,250 = N31,250/(N640/1US$) = $48.83/ton.

This analysis shows that a ton of PKS particles will cost about $49. This is in sharp contrast to the cost of alumina at $339.25 [106] and silicon carbide of $800–$2000 per ton [107]. That is, the cost of PKS reinforcing particles per ton is less than 15% of alumina, and between 2.5% and 6.1% of silicon carbide. Thus, it appears that there may be financial benefits in using this biodegradable material, as with others, such as listed in **Table 1**, as reinforcement materials for AMMCs.

3.10 Lightweighting index or effectiveness, L_x

The lightweighting index or effectiveness L_x is the dimensionless specific strength of the composite which we define as the ratio of specific strength of composite to the specific strength of the reference metal matrix, that is:

$$L_x = |\sigma| = \{(\sigma_c/\rho_c)/(\sigma_m/\rho_m)\} \tag{9}$$

The higher this value, the more strength per weight, the reinforcing material has over the reference metal matrix.

The table in the Appendix **Table A1** shows the L_x values for some of the references reviewed for which the density and tensile strength values for the reference metal matrix and composites were available.

Figure 5 shows the plots of lightweighting index L_x against wt.% composition of composites reinforced with egg shell ash [44] and breadfruit seed hull ash [33]. The figure shows that the breadfruit reinforcement seems to have a better strengthening power than the egg shell, going by these two test results [33, 44]. From the data points, the average lightweingthing index was found to be, $L_{xoEggShell}$ = 1.1487 and $L_{xoBreadfruit}$ = 1.1822.

3.11 Reinforcement performance index

Usually, it could be of interest to rate the reinforcing performance of different reinforcing materials from carefully conducted experiments or to rate the

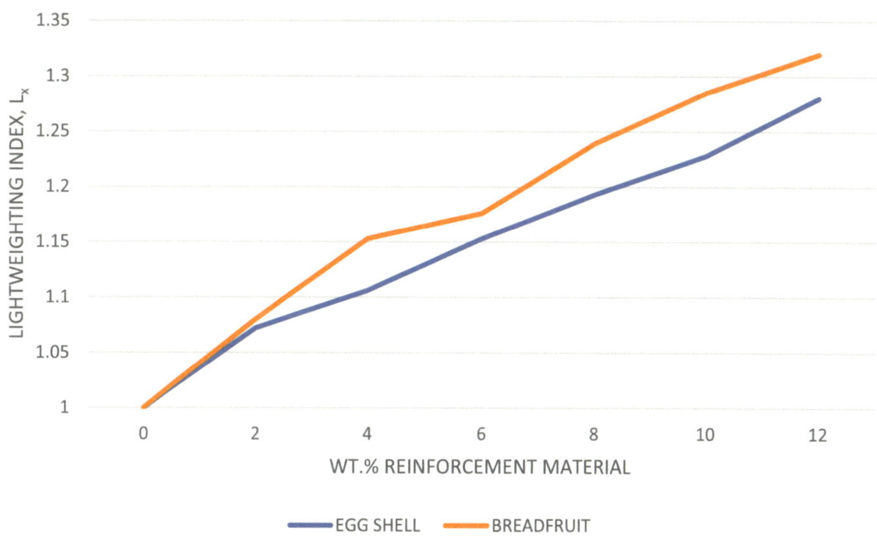

Figure 5.
Lightweighting index for two biodegradable green reinforcements

Property	Desirable change in property wrt reference metal matrix	Limiting condition
Density, ρ	Decrease	$\rho_c/\rho_m < 1$
Tensile strength, σ	Increase	$\sigma_c/\sigma_m > 1$
Hardness, H	Increase	$H_c/H_m > 1$
Impact energy, E	Increase	$E_c/E_m > 1$
Wear rate, W	Decrease	$W_c/W_m < 1$
Percent elongation, e	Increase	$e_c/e_m > 1$
Creep rate, c_r	Decrease	$c_{rc}/c_{rm} < 1$
Corrosion rate, K	Decrease	$K_c/K_m < 1$
Melting temperature, T	Increase	$T_c/T_m > 1$

Table 6.
Desirable changes of properties of composites for IC engines.

performance of various wt.% compositions for a particular reinforcement. We limit this analysis to the physical and mechanical properties of density, ρ, tensile strength σ, hardness H, impact energy E, percent elongation e, creep rate c_r, wear rate W, corrosion rate K and melting point T as may be applicable for IC engine parts.

Table 6 shows the desirable changes required of the composites with respect to the use of type of reinforcement and/or as the percentage weight of reinforcement increases.

From **Table 6**, the reinforcement performance index, P_x of the reinforcing material is maximized if

$$P_x = \frac{\frac{\sigma_c}{\sigma_m} \cdot \frac{H_c}{H_m} \cdot \frac{E_c}{E_m} \cdot \frac{e_c}{e_m} \cdot \frac{T_c}{T_m}}{\frac{\rho_c}{\rho_m} \cdot \frac{W_c}{W_m} \cdot \frac{K_c}{K_m} \cdot \frac{c_{rc}}{c_{rm}}} \tag{10}$$

If any property value is unavailable, that property ratio is set equal to 1 in Eq. (10).

Table 7 shows the computations of reinforcement performance index, P_x for breadfruit seed hull ash (BSHA) [33] and egg shell ash (ESA) [44] which do not have wear rate, elongation, creep rate, corrosion rate and melting point values. The table shows that the breadfruit seed hull ash has a higher performance index than the egg shell ash reinforcement.

3.12 The future of green materials as lightweighting reinforcement materials

There is a bright future for green materials as reinforcement materials for AMMCs as shown in **Table 1** for the following reasons:

i. As waste materials, they have the potential of being sources of low-cost reinforcements for AMMCs, as shown in the cost analysis in Section 3.9.

ii. Their lightweigthing potential is enormous because of their low densities compared to the aluminum metal matrix.

Reinforcement	Wt.%	ρ, g/cm³	ρ/ρ_m	σ, MPa	σ/σ_m	H, HRB	H/H_m	E, J	E/E_m	P_x
Breadfruit Seed hull Ash (BSHA)	0 (ref. metal matrix)	2.775	1.0000	172.5	1.0000	65	1.0000	12.6	1.0000	1.0000
	2	2.755	0.9928	185	1.0725	68	1.046	12	0.9523	1.0761
	4	2.735	0.9856	196	1.1362	72.5	1.115	11.2	0.8889	1.1426
	6	2.715	0.9784	198.5	1.1507	77.25	1.188	10.85	0.8611	1.2031
	8	2.675	0.9640	206	1.1942	81.25	1.25	10.4	0.8254	1.2781
	10	2.635	0.9495	210.5	1.2203	82.5	1.269	10.25	0.8135	1.3268
	12	2.620	0.9441	215	1.2464	85	1.308	10.15	0.8056	1.3911
	Average performance index, P_{xa}									**1.0597**
Egg shell ash (ESA)	0	2.775	1.000	98	1.000	58.5	1.000	14.5	1.000	1.0000
	2	2.675	0.964	101.25	1.033	62.5	1.068	12.6	0.869	0.9945
	4	2.65	0.955	103.5	1.056	64.5	1.103	12	0.828	1.0099
	6	2.615	0.942	106.5	1.087	67.5	1.154	11.8	0.814	1.0839
	8	2.6	0.937	109.5	1.117	68.5	1.171	9.8	0.676	0.9437
	10	2.56	0.923	111	1.133	69	1.179	9.6	0.662	0.9581
	12	2.5	0.901	113	1.153	74	1.265	8.3	0.572	0.9260
	Average performance index, P_{xa}									**0.9880**

Table 7.
Computation of reinforcement performance index, P_x for breadfruit seed hull ash (BSHA) [33] and egg shell ash (ESA) [44].

iii. Their relatively simple processing methods offer greater potential for use as AMMC reinforcement.

The following gaps exist for further research.

i. Carefully carried out experiments to determine the lightweighting effectiveness of these green reinforcing materials with respect to each other.

ii. Comparable tests on use of straight untreated particles, uncarbonized and carbonized ashed particles to determine the cost-effectiveness of these reinforcing particle modes.

4. Conclusion

This chapter has discussed the possibility of using biodegradable green materials as reinforcements for aluminum metal matrix composites. The following conclusions are drawn:

i. There is a wide range of biodegradable green materials that could be used as reinforcement materials for the lightweighting of aluminum alloys.

ii. The significant reduction in weight from use of these biodegradable green reinforcement materials depends on their smaller densities compared to the metal matrix.

iii. As waste materials, they provide a potential cheap source for reinforcing materials for AMMCs. The cost analysis shows that their cost per ton could be 15% of the cost of alumina and as low as 2.5% of cost of silicon carbide.

iv. Their use in composites gives lower densities, higher strength, and higher hardness over the reference metal matrix. However, their use decreases the impact strength of the composites.

v. Their strength values show that they can be used to make engine parts.

Acknowledgements

The author wishes to thank the Shell Petroleum Development Company of Nigeria (SPDC) and the Nigerian National Petroleum Corporation (NNPC) for support during his tenure as the NNPC/SPDC-JV Professor of Lightweight Automobile Engine Development at the Federal University of Petroleum Resources, Nigeria, from 2016 to 2020.

A. Appendix

DOI: http://dx.doi.org/10.5772/intechopen.108273

S/No	Green material	Ref	Work done	Results	Lightweighting parameter		
					Specific strength, $\sigma/\rho (N/mm^2)$		Lightweighting index, L_x, $\lvert\sigma\rvert = \lvert[\sigma_c/\rho_c]/[\sigma_m/\rho_m]\rvert$
					Ref. metal, σ_m/ρ_m	Best composite sample, σ_c/ρ_c	
1	Bagasse	[18]	A general review of reinforced AMMCs with little mention of published works on use of bagasse as reinforcement.	A feeble conclusion that bagasse can be used as reinforcement for AMMCs			
		[19]	Sugarcane Bagasse Ash from the co-generator boiler (SCBAB) of a Cuban sugar mill, was used as reinforcement for AMMC fabricated by employing powder metallurgy technique	The comparative study of the Al-SCBAB composite with respect to the reference metal matrix showed an increment in the hardness of the SCRAB reinforced AMMC over the reference matrix			
		[20]	The aging behavior for Al-Cu-Mg/Bagasse ash particulate composites with 2-10 wt% bagasse ash particles produced by double stir-casting method was investigated. The aging behavior of the solution and age-hardened reinforced composite was determined by use of hardness values and microstructural analysis	The reinforced composite exhibited an accelerated hardening response compared to the reference metal matrix			
		[21]	Fly ash and bagasse were used as reinforcement of Eutectic Al-Si alloy LM6 containing 10.58% Si to produce composites by liquid casting. Brinell hardness and ultimate tensile strength were determined	The result showed that 10 wt% fly ash +10 wt % bagasse ash can be used as reinforcement in aluminum composites especially for engine part			
		[22]	The properties of composites produced from an aluminum alloy Al-7%Si as matrix and two agro wastes Rice Husk Ash (RHA) and Bagasse Ash (BA) as reinforcement were compared	The results showed that BA had a better composite density lowering ability than RHA. However, the results show somewhat better improvement in mechanical properties with RHA addition	57.54	68.08	1.18
		[23]	Al6061 was reinforced with 4 wt.%, 8 wt.% and 12 wt.% bagasse ash to produce aluminum metal matrix composites. Microstructural and mechanical properties tests were carried out	The results showed that maximum tensile strength, minimum hardness, and maximum coefficient of friction occurred for 8 wt.% of bagasse ash content.			

S/No	Green material	Ref	Work done	Results	Lightweighting parameter		
					Specific strength, $\sigma/\rho (N/mm^2)$		Lightweighting index, L_x, $\lvert\sigma\rvert = \lvert[\sigma_c/\rho_c]/[\sigma_m/\rho_m]\rvert$
					Ref. metal, σ_m/ρ_m	Best composite sample, σ_c/ρ_c	
2	Bamboo leaf	[24]	Bamboo leaf ash (BLA) and alumina were used to reinforce Al-Mg-Si alloy matrix. The corrosion and wear behaviors of the resulting composites in 3.5% NaCl were investigated.	The corrosion resistance of the composites decreased with BLA addition. The composites at 2 and 3 wt.% of BLA showed superior wear resistance.			
		[25]	Bamboo leaf ash (BLA) and silicon carbide were used in various combined ratios to reinforce Al-Mg-Si alloy. Mechanical properties and corrosion behaviors of the composites were studied	The results indicated that all mechanical properties studied except fracture toughness decreased with increase in BLA content. Corrosion results were obtained for the composites in NaCl and H_2SO_4 media			
		[26]	Bamboo leaf ash (BLA) and alumina were used in various combined ratios to reinforce Al-Mg-Si alloy. Mechanical and microstructural properties of the composites were studied	The results indicated that all mechanical properties studied except fracture toughness decreased with increase in BLA content. The mechanical properties behavior were similar to that of silicon carbide/BLA reinforcement [6]. However, silicon carbide/BLA reinforcements tends to have a lower reduction on tensile strength than the alumina/BLA counterparts			
		[27]	Aluminum alloy AA6061 was reinforced with various combinations of silicon carbide and bamboo leaf ash (BLA), neem leaf ash (NLA) and tamarind leaf ash (TLA). Density and hardness measurements of the resulting composites were taken	The composite having only silicon carbide (1.5wt.%SiC) reinforcement had the highest density and highest hardness, while the composite with 0.75wt.%BLA and 0.75wt.%SiC had the least density and least hardness			
		[28]	This is an extension of Ref. [27] by one of the authors where more tests were carried out.	The results showed that BLA reinforcement (0.75wt.%BLA and 0.75wt.%SiC) had the highest tensile strength, highest hardness and least density and least wear. Some of these results are at variance with [29]	45.30	58.79	1.30

S/No	Green material	Ref	Work done	Results	Lightweighting parameter		
					Specific strength, $\sigma/\rho (N/mm^2)$		Lightweighting index, L_x, $\lvert\sigma\rvert = \lvert[\sigma_c/\rho_c]/[\sigma_m/\rho_m]\rvert$
					Ref. metal, σ_m/ρ_m	Best composite sample, σ_c/ρ_c	
3	Bone	[30]	Al6063 alloy was reinforced with alumina and chicken bone ash to produce five composite samples each having a total of 8 wt.% content. In each of the five compositions, bone ash was 2, 3, 4, 5 and 6 wt.% while alumina took up the balance in 8 wt.%. Machining tests were carried out on the composites to determine surface roughness	The results showed that the 2 wt.%Alumina and 6 wt.% chicken bone ash best surface roughness compared to the others			
		[31]	Al6063 alloy was reinforced with calcined cow bone (CB) particles and crushed periwinkle shell (PS) particles separately with 1, 2, 4, 6, 8 and 10 wt.% to produce six compositions each. They were used to make knee braces. Physical and mechanical tests were carried out on them.	The periwinkle shell reinforcement had the best mechanical properties and smaller density at all compositions compared to the cow bone. For the cow bone reinforcement, the density decreased and tensile strength increased as its percent content increased, while it maximum hardness occurred at 6 wt.%	47.62	51.57 (CB)	1.09
		[32]	This paper claims that aluminum 8081 alloy was reinforced with chicken bone particles.	The presentation of results was not clear. However, the authors claimed that "it was found that the Chicken Bone Ash contents were able to improve the Wear rate of aluminum metal matrix composite compared to the aluminum-8011".			
4	Breadfruit seed hull	[33]	Al-Si-Fe alloy was reinforced with breadfruit seed hull ash in weight percentages of 0–12 in steps of 2 wt.% to produce composites. The mechanical and microstructural properties were determined	The results showed that density decreased, tensile strength and hardness as breadfruit content in composites increased	62.16	81.90	1.32

S/No	Green material	Ref	Work done	Results	Lightweighting parameter		
					Specific strength, $\sigma/\rho (N/mm^2)$		Lightweighting index, L_x, $\lvert\sigma\rvert = \lvert[\sigma_c/\rho_c]/[\sigma_m/\rho_m]\rvert$
					Ref. metal, σ_m/ρ_m	Best composite sample, σ_c/ρ_c	
5	Coconut shell	[34]	Aluminum alloy 6063 was reinforced with coconut shell ash (CSA). The properties of the composites were measured, and the corrosion properties in 0.3 M H2SO4 and 3.5 wt% NaCl solutions were also determined	The results showed that density decreased, tensile strength and hardness increased as CSA content increased. Composites with 12 wt.% CSA had the least corrosion in both 0.3 M H_2SO_4 and 3.5 wt.% NaCl solutions	—	—	
		[35]	Coconut shell particles (CSP) of sizes 50, 75, 150, 212, and 300 μ were used to reinforce recycled waste aluminum cans at 5 wt.% and 10 wt.% to produce composites. The mechanical and wear properties were determined	The tensile strength and hardness increased and impact strength decreased as CSP content increased. As CSP contentment increases, the wear rate becomes significant at bearing load over 14.21 N	—	—	
		[36]	99.70% purity aluminum was reinforced with coconut shell ash (CSA) at 3 – 15 wt.% in steps of 3 wt.%. The microstructure and mechanical properties were determined	Hardness increased as CSA content increased, while maximum tensile strength occurred at 9 wt.% of CSA	—	—	
		[37]	Density, particle size, refractoriness, SEM, XRD, XRF and FTIR spectroscopic methods were used for the characterization of the coconut shell ash to determine its suitability as a reinforcement material in metal matrix composites.	From the microstructural analysis of the composites, the authors conclude: "The coconut shell ash can withstand a temperature of up to 1500oC with a density of 2.05 g/cm3. That means this ash can be used in production of light weight MMCs component with good thermal resistance"	—	—	
		[29]	Pure aluminum was reinforced with 5wt.%SiC for three sample compositions of 3 wt.%, 5 wt.% and 10 wt.% coconut shell ash (CSA). The composites were tested for density, hardness, tensile strength and impact energy.	Density and impact strength decreased, while hardness and tensile strength increased with CSA content.	19.64	69.44	3.54
		[38]	99.3% purity aluminum was reinforced with finely ground coconut shell particles in amounts of 2 wt.% to 10 wt.% in steps of 2 wt.%. Microstructural and mechanical property studies were carried out.	Tensile strength and hardness increased while density and impact energy decreased as coconut shell particle content increased.	38.89	81.82	2.10

S/No	Green material	Ref	Work done	Results	Lightweighting parameter		
					Specific strength, $\sigma/\rho (N/mm^2)$		Lightweighting index, L_x, $\lvert\sigma\rvert = \lvert[\sigma_c/\rho_c]/[\sigma_m/\rho_m]\rvert$
					Ref. metal, σ_m/ρ_m	Best composite sample, σ_c/ρ_c	
		[39]	Paper claims that Al1100 was reinforced with 15% coconut shell ash to produce composites for machining to determine wear characteristics using Taguchi experimental design approach.	The relevant result was that the coconut shell ash reinforced composite had higher values of tensile strength and hardness above the reference alloy Al1100			
		[40]	Waste aluminum cans alloy was reinforced with coconut shell ash (CSA) (150 μm) and graphite (150 μm) particles. All five composition samples had 2 wt.% graphite while CSA content was 0, 2, 4, 6 and 8 wt.%. Microstructural, physical and mechanical tests were carried out on the produced composites	The results showed that density and impact strength decreased, while tensile strength and hardness increased as CSA content increased. Graphite increased/decreased the properties of the aluminum alloy, such that addition of CSA gave greater increases/decreases	10.98	22.87	2.08
6	Corncob	[41]	Aluminum alloy 8011 was reinforced separately with cow horn and corncob with contents from 5 wt.% - 20 wt.% in steps of 5 wt.% to produce aluminum metal matrix composites. Microstructural and mechanical tests were carried out	Maximum tensile strength and maximum elongation occurred at 15 wt.%. Hardness decreased as corncob content increased			
		[42]	Al-Mg-Si alloy matrix was reinforced with various percentage weights of SiC and corncob ash to produce aluminum metal matrix composites. Microstructural and mechanical tests were carried out.	Density, hardness, elongation and tensile strength decreased and fracture toughness increased for composites as corncob content increased in the composites			
		[43]	Aluminum alloy 6063 was reinforced with corncob ash in weight percentages of 2.5% - 15% in steps of 2.5%. microstructural and mechanical property tests were carried out	The results showed that density, tensile stress, impact strength, and wear decreased; hardness increased as corncob ash content increased in the composites.			

S/No	Green material	Ref	Work done	Results	Lightweighting parameter		
					Specific strength, $\sigma/\rho (N/mm^2)$		Lightweighting index, L_x, $\lvert\sigma\rvert = \lvert[\sigma_c/\rho_c]/[\sigma_m/\rho_m]\rvert$
					Ref. metal, σ_m/ρ_m	Best composite sample, σ_c/ρ_c	
7	Cow horn	[41]	Aluminum alloy 8011 was reinforced separately with cow horn and corncob with contents from 5 wt.% - 20 wt.% in steps of 5 wt.% to produce aluminum metal matrix composites. Microstructural and mechanical tests were carried out	Tensile strength increased and elongation decreased as cow horn content increased. Maximum hardness occurred at 15 wt.%			
8	Egg shell	[44]	Eggshell particles (ES) (uncarbonized and carbonized) were used to reinforce Al–Cu–Mg (A2009 alloy) to produce composites. A total of 2–12 wt.% ES particles were added in steps of 2 wt.%. Microstructural and mechanical properties were studied	For both carbonized and uncarbonized egg shell particles, density and impact energy decreased, tensile strength and hardness increased for increase in egg shell particles in composites. For strength and hardness the carbonized values were higher than the uncarbonized; while for density and impact energy, the carbonized had lower values.	35.50	45.20 (carbonized) 41.36 (uncarbonized)	1.27 (carbonized) 1.17 (uncarbonized)
		[45]	Al6061 was reinforced with eggshell uncarbonized powder from 2 wt.% - 10 wt.% in steps of 2 wt.%. microstructural and mechanical properties tests were carried out	Density of composites decreased as eggshell powder content increased. For both tensile strength and hardness, their maximum values occurred at 4 wt.% of eggshell powder	37.65	45.34	1.20
		[46]	AA2014 aluminum alloy was reinforced with carbonized and uncarbonized eggshell particles. Calcium carbonate was also used as reinforcement separately. Microstructural and mechanical properties tests were carried out	Tensile strength, hardness and fatigue strength increased as eggshell content increased up to 12.5 wt.% in AA2014 for both carbonized and uncarbonized reinforced composites. Toughness, ductility and corrosion rate decreased as eggshell increased up to 12.5 wt.% for both carbonized and uncarbonized reinforced composites.	—	—	

S/No	Green material	Ref	Work done	Results	Lightweighting parameter		
					Specific strength, $\sigma/\rho (N/mm^2)$		Lightweighting index, L_x, $\lvert\sigma\rvert = \lvert[\sigma_c/\rho_c]/[\sigma_m/\rho_m]\rvert$
					Ref. metal, σ_m/ρ_m	Best composite sample, σ_c/ρ_c	
		[47]	Eggshell particles (carbonized and uncarbonized), SiC particles,, and $CaCO_3$ powder were used to reinforce AA2014 aluminum alloy separately. Test properties were carried out on the composites produced.	By addition of SiC particles up to 10 wt.% and waste eggshell particles up to 12.5 wt.%, the tensile strength, hardness, and fatigue strength increased. Toughness and ductility decreased by the addition of SiC and eggshell particles. Corrosion rate decreased by the addition of SiC particle up to 7.5 wt.% and eggshell particles up to 12.5 wt.%. Hardness and heat-treatable properties are improved after the addition of SiC reinforcement particles as compared to eggshell particles			
		[48]	Uncarbonized eggshell particle was used to reinforce AA2014 aluminum alloy. Property tests were carried out on the derived composites. Process parameters of the electromagnetic stir casting process used were varied to find their influence on the properties of AA201/uncarbonized eggshell particles	Reinforcement parameters of preheat temperature, stirring current, stirring time, matrix pouring temperature, and reinforcement weight percentage affected the tensile strength of the composites. The maximum tensile strength of 287.194 MPa was found for reinforcement preheat temperature 537.87 °C, stirring current 12 A, stirring time 179.9 sec, matrix pouring temperature 726.8 ° C, and reinforcement weight percentage of 12.46.			
		[49]	Al6061 powder was reinforced with 5 wt%, 10 wt.% and 15 wt.% of eggshell powder and composites were produced by powder metallurgy technique. Microstructural, physical and mechanical properties test were carried out.	Hardness increased, while density and electrical conductivity decreased as eggshell content increased in the composites			

S/No	Green material	Ref	Work done	Results	Lightweighting parameter		
					Specific strength, $\sigma/\rho(N/mm^2)$		Lightweighting index, L_x, $\lvert\sigma\rvert = \lvert[\sigma_c/\rho_c]/[\sigma_m/\rho_m]\rvert$
					Ref. metal, σ_m/ρ_m	Best composite sample, σ_c/ρ_c	
9	Groundnut shell	[50]	Zn-27Al alloy was reinforced with SiC and groundnut shell ash. Corrosion, microstructural and mechanical property tests were carried out on the derived composites	Hardness and ultimate tensile strength of the hybrid composites decreased with increase in GSA content; while the fracture toughness of the hybrid composites increased with GSA content increase. In 3.5% NaCl solution, the composites were resistant to corrosion, while in 0.3 M H2SO4 solution, the composites were not as resistant to corrosion			
		[51]	Al6063 aluminum alloy was reinforced with groundnut shell ash in weight percents of 3, 6, 9, and 12 wt.%. Microstructural and mechanical property tests were carried out on the derived composites	As groundnut shell ash increased, tensile strength increased on toa maximum at 9 wt.% of reinforcement, hardness and compressive strength increased, while percent elongation and impact energy decreased			
		[52]	Al-Mg-Si aluminum alloy was reinforced with silicon carbide and groundnut shell ash to produce composites. Corrosion studies were carried out on the composites	The composites were resistant to corrosion in 3.5% NaCl solution, but more susceptible to corrosion in 0.3 M H2SO4 solution.			
		[53]	Al-Mg-Si aluminum alloy was reinforced with silicon carbide and groundnut shell ash to produce composites. Microstructural, mechanical and fracture property studies were carried out	Hardness and tensile strength increased with increasing GSA content but the strength and hardness dropped slightly with an increase in GSA content. Fracture toughness improved with increase in GSA content.			
10	Lemon Grass	[54]	Aluminum alloy Al6061 was reinforced with 3 wt.%, 5 wt.% and 7 wt.% lemon grass ash. Microstructural and mechanical properties of the derived composites were determined	It was reported that the there was linear increase of tensile strength and hardness with increase in lemon grass ash content			

S/No	Green material	Ref	Work done	Results	Lightweighting parameter		
					Specific strength, $\sigma/\rho (N/mm^2)$		Lightweighting index, L_x, $\lvert\sigma\rvert = \lvert[\sigma_c/\rho_c]/[\sigma_m/\rho_m]\rvert$
					Ref. metal, σ_m/ρ_m	Best composite sample, σ_c/ρ_c	
11	Maize Stalk	[55]	Al-Si-Mg alloy was reinforced with maize stalk ash at weight percentages of 2, 4, 6, 8 and 10%. Microstructural and mechanical properties of composites were determined				
12	Mango Seed Shell	[56]	Al-Si-Mg alloy was reinforced with mango seed shell particle at contents of 5 wt.%, 10 wt.%, 15 wt.% and 20 wt.%. Microstructural, wear and some mechanical property tests were carried out on the derived composites	Maximum hardness and maximum impact energy of 43.2 HV and 2.44 J respectively were obtained at 15 wt.% mango seed shell content. Also, at 15 wt.%, wear resistance improved on that of the reference metal matrix			
		[57]	Essentially same work as reference [56] except that the optimum 15 wt.% mango seed shell composite was used to produce a motorcycle hub	The production of the motorcycle hub with the 15 wt.% mango seed shell particles was successful			
		[58]	Taguchi experimental design technique was used to optimize the production parameters of Al-Si-Mg alloy reinforced with mango seed shell ash	Optimal wear rate of the mango seed shell ash reinforced-reinforced Al–Si–Mg composite was found to be 0.001517 $mm^3/N/m$ at stirring time, processing temperature, MSSA content, and particle size of 60 s, 720°C, 20%, and 25 μm respectively.			
13	Marula	[59]	Al-Mg-Si alloy was reinforced with carbonized marula seed cake in percent weights from 0–14% in steps of 2%. It was used for the production of brake pads. Physical and mechanical properties of the composites were carried out	The results showed that tensile strength and hardness increased while the density, percent elongation and impact energy decreased as the carbonized marula seed cake content increased			

S/No	Green material	Ref	Work done	Results	Lightweighting parameter		
					Specific strength, $\sigma/\rho (N/mm^2)$		Lightweighting index, L_x, $\lvert\sigma\rvert = \lvert[\sigma_c/\rho_c]/[\sigma_m/\rho_m]\rvert$
					Ref. metal, σ_m/ρ_m	Best composite sample, σ_c/ρ_c	
14	Neem Leaf	[60]	Al6061 T6 alloy was reinforced with three reinforcements: 20%SiC,!0%SiC +10%Fly Ash, and 10%SiC +10% Neem leaf Ash to produce three metal composites. Machine lever was produced with the composite	The tests showed that the neem leaf ash reinforcement produced the best property values of highest hardness, highest tensile strength and least wear.			
		[28]	This is an extension of Ref [27] by one of the authors where more tests were carried out.	The composite with 0.75wt.%NLA/0.75wt.% SiC reinforcement had the intermediate tensile strength, intermediate yield strength, intermediate elongation and intermediate wear values. The bamboo leaf ash had the best performance			
		[27]	Aluminum alloy AA6061 was reinforced with various combinations of silicon carbide and bamboo leaf ash (BLA), neem leaf ash INLA) and tamarind leaf ash (TLA). Density and hardness measurements of the resulting composites were taken	The composite having only silicon carbide (1.5wt.%SiC) reinforcement had the highest density and highest hardness, while the composite with 0.75 wt.% neem leaf ash and 0.75wt.%SiC had intermediate values for density and hardness			
15	Palm Kernel Shell (PKS)	[61]	Aluminum alloy 6063 was reinforced with varying weight percentage of 212 μm palm kernel shell (PKS) particles (2.5%, 5%, 7.5%, 10%, 12.5% and 15%). Microstructural, chemical and physical property tests were carried out on the composites produced	Density of composites decreased as PKS content increased while the porosity increased though within acceptable values			
		[62]	Disused engine block aluminum was reinforced with PKS ash particles in weight percentages of 5, 10 and 15. One half of PKS ash was treated with 1 M NaOH, while the other half remained untreated, thus, giving three sample composition each for the treated and untreated. Microstructural, physical and mechanical property tests were carried out on the composites	The mechanical property test results did not follow any discernable pattern, apart from impact energy for treated PKS compositions which showed a trend of increasing impact energy as PKS content increased.			

S/No	Green material	Ref	Work done	Results	Lightweighting parameter		
					Specific strength, $\sigma/\rho(N/mm^2)$		Lightweighting index, L_x, $\|\sigma\|$ = $\|[\sigma_c/\rho_c]/[\sigma_m/\rho_m]\|$
					Ref. metal, σ_m/ρ_m	Best composite sample, σ_c/ρ_c	
		[63]	Aluminum alloy A356 was reinforced with PKS nanoparticles at weight percentages of 1, 2, 3, and 4. Microstructural, physical and mechanical properties were determined.	The results showed improvements of 30.47%, 41.91%, 49.52%, 40.90% and 65.09% on hardness values, tensile, yield strength, % elongation and impact energy at 4 wt% PKS over reference metal matrix.	40.38	66.75	1.65
		[64]	Aliminum AA6063 was reinforced with different mass fractions of palm kernel shell (PKS) particles (0, 2.5, 5, 7.5, 10, 12.5 & 15 wt %). Microstructural and wear tests were carried out on the composites produced	Better wear properties were obtained by the 10 wt% PKS reinforcement.			
		[65]	Zn-Al alloy was reinforced with 5 wt% SiC added with 0.2%, 0.4%, 0.6%, 0.8% and 1.0 wt % PKS ash particles. Microstructural and mechanical properties were determined.	The results showed that tensile strength and hardness of composites increased as percent weight of PKS ash increased in the composites			
		[66]	Commercially pure aluminum was reinforced with combined PKS and periwinkle shell (PS) particles with the optimum percent weights of PKS and PS determined by central composite design (CCD) of response surface methodology (RSM)	The fabricated composite had significantly improved properties over the commercially pure aluminum	27.31	36.21	1.33
		[5]	Same conditions as Ref. [66] but used to produce an engine block for a brush cutter	The test results of the produced engine block showed that the composite could be used for manufacture of engine parts	27.31	36.21	1.33
16	Palm Sprout Shell	[67]/ [68]	[67] reported that [68] reinforced alumirum alloy with palm sprout shell ash in weight fractions of 1%,2%, 3%. Stir casting was used to produce the composites. (The reference could not be accessed)	It was stated by [67] that a maximum hardness of 100BHN was obtained for 3% weight fraction of palm sprout shell ash			

S/No	Green material	Ref	Work done	Results	Lightweighting parameter		
					Specific strength, $\sigma/\rho(N/mm^2)$		Lightweighting index, L_x, $\|\sigma\| = \|[\sigma_c/\rho_c]/[\sigma_m/\rho_m]\|$
					Ref. metal, σ_m/ρ_m	Best composite sample, σ_c/ρ_c	
17	Periwinkle Shell (PS)	[69]	Aluminum 6063 was reinforced with periwinkle shell particle sizes of 75 μm and 150 μm at weight percentages 1, 5, 10 and 15. The mechanical properties and microstructures of the composites were determined	The mechanical results do not present a discernable pattern of variation except for 75 μm particle composites for tensile strength and hardness where these values increase with reinforcement content			
		[70]	Similar to [71] except for determination of density	Similar results to [5] except that density results follow a discernable pattern. Density of composites decreased as reinforcement content increased, and that 150 μm PS particles had less density than corresponding composite densities of 75 μm as the latter had greater porosities.			
		[72]	Al-3.7%Cu-1.4%Mg was reinforced with periwinkle shell ash of weight percentages of 0–30% in steps 5%. The derived composites were tested for microstructure, and physical and mechanical properties	The results of the mechanical properties tests showed that, the addition of periwinkle ash increased the hardness, decreased the density and also decreased the impact energy of the composites produced for all additions. The tensile strength increased from 153.75 N/mm2 at 0 wt% to 202.45 N/mm2at 25 wt% PS.			
		[5]	Same as for PKS				
		[66]	Same as for PKS				
		[31]	Al6063 alloy was reinforced with calcined cow bone (CB) particles and crushed periwinkle shell (PS) particles separately with 1, 2, 4, 6, 8 and 10 wt.% to produce six compositions each. They were used to make knee braces. Physical and mechanical tests were carried out on them.	For both CB and PS reinforcements, density decreased, tensile strength and hardness increased compared to the reference alloy as percent reinforcement increased. PS reinforced composites had superior property values than CB	47.62	54.17 (PS)	1.14

S/No	Green material	Ref	Work done	Results	Lightweighting parameter		
					Specific strength, σ/ρ (N/mm^2)		Lightweighting index, L_x, $\lvert\sigma\rvert = \lvert[\sigma_c/\rho_c]/[\sigma_m/\rho_m]\rvert$
					Ref. metal, σ_m/ρ_m	Best composite sample, σ_c/ρ_c	
18	Rice Husk	[71]	Reviews the works done in using rice husk as a reinforcement in AMMCs	The review concludes that with appropriate processing method and choice of process parameters, use of carbonized rice husk could further enhance the mechanical properties and widen the scope of usage in automotive applications			
		[21]	The properties of composites produced from an aluminum alloyAl-7%Si as matrix and two agro wastes Rice Husk Ash (RHA) and Bagasse Ash (BA) as reinforcement were compared	The results showed that BA had a better composite density lowering ability than RHA. However, the results show somewhat better improvement in mechanical properties with RHA addition			
		[73]	Aluminum alloy, A356.2 was reinforced with 2 wt.%, 4 wt.% and 8 wt.% of rice husk ash (RHA) particles. Microstructural and mechanical tests were carried out on the derived composites	Results showed that tensile strength and harness increased while percent elongation decreased as reinforcement content increased in the composite			
		[74]	Aluminum alloy A356.2 was reinforced with RHA in percent weights of 4, 6 and 8%, and composites were made by vortex casting. Microstructural, physical and mechanical tests were carried out on the composites	Results showed that composites' tensile strength and hardness increased while density of decreased as RHA content increased	100.54	121.40	1.21
		[75]	Al-Mg-Si alloy was reinforced with RHA and SiC mixed in weight ratios 0:1, 1:3, 1:1, 3:1, and 1:0. They were utilized to prepare 5, 7.5 and 10 wt% of the reinforcing phase using two-step stir casting method. Mechanical properties were investigated	The results showed a general increase in tensile strength, specific strength and hardness, and genera decrease in density, impact toughness and ductility with increase in weight percent of RHA/SiC. The composites with composition 25% RHA: 75% SiC offers comparable specific strength values with the SiC single reinforced grades of the composite.	≃ 38	≃ 60	1.58

S/No	Green material	Ref	Work done	Results	Lightweighting parameter		
					Specific strength, σ/ρ(N/mm^2)		Lightweighting index, L_x, $\lvert\sigma\rvert = \lvert[\sigma_c/\rho_c]/[\sigma_m/\rho_m]\rvert$
					Ref. metal, σ_m/ρ_m	Best composite sample, σ_c/ρ_c	
		[76]	Al-Mg-Si alloy was reinforced with RHA and SiC mixed in weight ratios 0:1, 1:3, 1:1, 3:1, and 1:0. They were utilized to prepare 5, 7.5 and 10 wt% of the reinforcing phase using two-step stir casting method. Corrosion and wear behavior were investigated.	The results showed that in general, hybrid reinforcement of RHA and SiC resulted in improved corrosion resistance of the composites in 3.5% NaCl solution. The coefficient of friction/wear resistance of the hybrid composites were comparable to that of the Al–Mg–Si alloy matrix reinforced with only SiC.			
		[77]	Al–Mg–Si alloy matrix was reinforced with rice husk ash (RHA) and alumina. Alumina added with 2, 3, and 4 wt.% RHA were utilized to prepare 10 wt.% of the reinforcing phase. Corrosion and wear behavior of the composites were investigated	The corrosion resistance of the hybrid composites in 3.5% NaCl solution was less than that of the single reinforced Al–Mg–Si/10 wt.% Al2O3 composite. The corrosion rates increased with increase in wt.% RHA. The coefficient of friction/wear rate of the composites increased with increase in RHA wt.%.			
		[78]	Waste aluminum cans were reinforced rice husk ash particles of 150, 300 and 600 μm in weight percentages of 5, 10 and 15. Physical, mechanical and microstructural properties were investigated	The results shows that density decreases with increase percent weight of RHA, and suggest that for same percent weight of RHA, density decreases as RHA particles reduce in size. Other results are: tensile strength and hardness increase; and impact strength decrease with reduction in RHA particle size.	51.93	70.88	1.36
		[79]	A6061 aluminum alloy was reinforced with silicon carbide and rice husk ash for each composite made at 8 wt.% for each. The ratios of RHA and SiC in the composite were 1:4, 2:3 and 0:1 in the 8 wt.%. The microstructure, density, porosity and mechanical properties of the composites were determined	The results showed that the less dense Al/RHA/SiC hybrid composites have estimated percent porosity levels as low as <2.86% porosity.			

DOI: http://dx.doi.org/10.5772/intechopen.108273

S/No	Green material	Ref	Work done	Results	Lightweighting parameter		
					Specific strength, $\sigma/\rho (N/mm^2)$		Lightweighting index, L_x, $\lvert\sigma\rvert = \lvert[\sigma_c/\rho_c]/[\sigma_m/\rho_m]\rvert$
					Ref. metal, σ_m/ρ_m	Best composite sample, σ_c/ρ_c	
		[80]	The paper states that pure aluminum was reinforced with constant 5 wt.% of rice husk ash and 2 wt.%, 4 wt.% and 6 wt.% of silicon carbide to produce hybrid composites. Microstructural, harness and strength tests were carried out	The results showed that as percent weight of silicon carbide increased, the hardness increased, while the compressive strength reached a maximum value at 4 wt.% of SiC			
		[22]	The properties of composites produced from an aluminum alloyAl-7%Si as matrix and two agro wastes Rice Husk Ash (RHA) and Bagasse Ash (BA) as reinforcement were compared	The results showed that BA had a better composite density lowering ability than RHA. However, the results show somewhat better improvement in mechanical properties with RHA addition			
		[81]	86.75Al alloy obtained from meting automobile parts, roofing sheet, can, etc. was reinforced with rice husk ash of content 0 vol.% - 30 vol.% in steps of 5 vol.% to produce composite samples. Physical and mechanical tests were carried out	The results showed that density decreased as RHA content increased, maximum tensile strength, maximum impact strength, maximum hardness occurred at 10 vol.%, 10 vol.% and 25 vol.% of RHA content respectively	57.54	65.34	1.14
		[82]	AluminumA356.2 alloy was hybrid reinforced with 4wt.%rice husk ash (RHA) + 6 wt.% fly ash (FA), 5wt.%RHA + 5wt.%FA and 6 wt.% RHA + 4wt.%FA to produce composites. Mechanical tests were carried out on the composites	The results showed that hardness, tensile strength, compressive strength and impact strength occurred for 6 wt.%RHA + 4wt.%FA, while elongation decreased somewhat as RHA content increased			
		[83]	Aluminum AlSi10Mg alloy was reinforced with RHA in percent weights 3, 6, 9 and 12 to produce composites. They were subjected to microstructural and mechanical tests	The results showed that tensile strength, compressive strength and harness increased while percent elongation decreased as RHA content increased			
		[84]	Eutectic Al-Si alloy LM6 was reinforced separately with RHA and FA. However, the percent weights of the reinforcements were not given. Mechanical property and machining tests were carried out.	The results showed that FA reinforced composite samples had better mechanical properties than the RHA. It was also shown that the FA composite gave lower cutting force and better surface finish			

S/No	Green material	Ref	Work done	Results	Lightweighting parameter		
					Specific strength, $\sigma/\rho (N/mm^2)$		Lightweighting index, L_x, $\lvert\sigma\rvert = \lvert[\sigma_c/\rho_c]/[\sigma_m/\rho_m]\rvert$
					Ref. metal, σ_m/ρ_m	Best composite sample, σ_c/ρ_c	
19	Snail shell	[85]	Aluminum alloy obtained from discarded pistons and roofing sheets was reinforced with snail shell particles of percent weight s from 16 to 48 wt.%. Microstructural and mechanical tests were carried out on the derived composites.	The results showed that, at 48 wt.% and 600 μm particle size, the tensile strength and hardness are maximized to 236 MPa and 48.3 HRF respectively compared to the tensile strength of 92.4 MPa and hardness of 29.2 HRF for the unalloyed samples.			
20	Tamarind leaf	[27]	Aluminum alloy AA6061 was reinforced with various combinations of silicon carbide and bamboo leaf ash (BLA), neem leaf ash INLA) and tamarind leaf ash (TLA). Density and hardness measurements of the resulting composites were taken	The composite having only silicon carbide (1.5wt.%SiC) reinforcement had the highest density and highest hardness, while the composite with 0.75 wt.% tamarind leaf ash and 0.75wt.%SiC had the highest density and least porosity			
		[28]	This is an extension of Ref [27] by one of the authors where more tests were carried out.	The composite with 0.75wt.%TLA/0.75wt.% SiC reinforcement generally had performance values next after NLA. BLA had the best performance values			

Table A1.
Summary of work done in using green materials as reinforcement of AMMCs.

Author details

Akaehomen O. Akii Ibhadode
Department of Production Engineering, University of Benin, Benin City, Nigeria

*Address all correspondence to: ibhadode@uniben.edu

References

[1] Anon. Lightweight Materials for Cas and Trucks, Vehicles Technologies Office, Office of Energy Efficiency & Renewable Energy, U. S. Department of Energy. Available from: https://www.energy.gov/eere/vehicles/lightweight-materials-cars-and-trucks. [Accessed: 20 July 2022]

[2] Järvikivi M, The Road to Automotive Lightweighting, Industrial Heating. Available from: https://www.industrialheating.com/articles/96356-the-road-to-automotive-lightweighting. [Accessed 20 July 2022]

[3] Prucz JC, Shoukry SN, William GW, Shoukry MS. Lightweight Composite Materials for Heavy Duty Vehicles — Final Report. US Department of Energy, Office of Energy Efficiency and Renewable Energy, National Energy Technology Laboratory; 2013 Contract No. DE-FC26-08NT02353, Department of Mechanical and Aerospace Engineering College of Engineering and Mineral Resources West Virginia University

[4] Isenstadt A, German J, Bubna P, Wiseman M, Venkatakrishnan U, Abbasov L, et al. Lightweighting technology development and trends in U.S. passenger vehicles, Working Paper 2016-25, The International Council on Clean Transportation (ICCT), 16 December 2016. Available from: https://theicct.org/wp-content/uploads/2021/06/ICCT_PVtech_lightweighting_wp2016-25.pdf

[5] Ibhadode AOA, Ebhojiaye RS. A New Lightweight Material for Possible Engine Parts Manufacture [Online First]. IntechOpen. 2018. DOI: 10.5772/intechopen.82268. Available from: https://www.intechopen.com/online-first/a-new-lightweight-material-for-possible-engine-parts-manufacture

[6] Date PP. Lightweighting of IC Engines. In: Efficient Manufacturing (A Web Magazine of industr.com) January 23, 2020. 2020. Available from: https://www.industr.com/en/lightweighting-of-ic-engines-2467925

[7] de Araujo AN. Automotive Lightweighting: Strategies for Improving Fuel Efficiency. 2021. Available from: https://www.prescouter.com/2021/09/automotive-lightweighting-strategies-for-improving-fuel-efficiency/

[8] Babalola PO, Bolu CA, Inegbenebor AO, Odunfa KM. Development of aluminium matrix composites: A review. Online International Journal of Engineering and Technology Research. 2014, 2014;**2**:1-11 ISSN 2346-7452

[9] Soboyejo W. Mechanical Properties of Engineered Materials. New York: Marcel Dekker Inc.; 2003

[10] Miracle DB. Metal matrix composites-from science to technological significance. Composites Science and Technology. 2005; **65**(15–16):2526-2540

[11] Anon. Circular Economy and the Stockholm Convention Division, Department of Environment, United Nations Industrial Development Organization (UNIDO). Vienna, Austria. 2019

[12] Anon. The Circular Economy is Nature's Equivalent of 'Living Within Your Means', Circular Economy. Available from: https://www.circle-economy.com/circular-economy/what-is-the-circular-economy. [Accessed: 20 July 2022]

[13] Anon. Green Industrial Policy: Concept, Policies, Country Experiences, UN Environment, Nairobi, Kenya. 2017

[14] Anon. Going green, Middletown Thrall Library Presents. Available from: https://www.thrall.org/special/goinggreen.html. [Accessed: 20 July 2022]

[15] Joseph OO, Babaremu K. Agricultural waste as a reinforcement particulate for aluminum metal matrix composite (AMMCs): A review. Fibers. 2019;7(33):1-9

[16] Kristoferson LA, Bokalders V. Renewable Energy Technologies: Their Applications in Developing Countries. Elsevier Science. 2013

[17] Anon. Green Materials. NEXT.CC. Available from: https://www.next.cc/journey/discovery/green-materials [Accessed: 4 August 2022]

[18] Srivastava N, Dwivedi SP. Development of green hybrid metal matrix composite using agricultural waste bagasse as reinforcement—A review. In: IOP Conf. Series: Materials Science and Engineering. Vol. 691. IOP Publishing; Greater Noida, India: GL Bajaj Institute of Technology and Management; 2019. p. 012051. DOI: 10.1088/1757-899X/691/1/012051

[19] Hern'andez-Ruiz JE, Pino-Rivero L, Villar-Coci E. Aluminium matrix composite with sugarcane bagasse ash as reinforcement material. Revista Cubana de Física. 2019;**36**:55

[20] Aigbodion VS, Hassan SB, Dauda ET, Mohammed RA. Experimental study of ageing behaviour of Al-Cu-Mg/bagasse ash particulate composites. Tribology in industry. 2011;**33**(1):28-35

[21] Khan MZ, Anas M, Hussain A, Haque MIL, Rasheed K. Effect on mechanical properties of aluminium alloy composites on adding ash as reinforcement material. Journal of Metals, Materials and Minerals. 2015;**25**(2):1-7

[22] Usman AM, Raji A, Hassan MA, Waziri NH. A comparative study on the properties of Al-7%Si-rice husk ash and Al-7%Si-bagasse ash composites produced by stir casting. The International Journal Of Engineering And Science (IJES). 2014;**3**(8):01-07

[23] Anwesh K. Virkunwar, Shouvik Ghosh, Ranjan Basak, Study of mechanical and tribological characteristics of aluminium alloy reinforced with sugarcane bagasse ash, An International Conference on Tribology, TRIBOINDIA-2018, 2018, VJTI, Mumbai, India

[24] Alaneme KK, Olubambi PA, Afolabi AS, Bodurin MO. Corrosion and tribological studies of bamboo leaf ash and alumina reinforced Al-Mg-Si alloy matrix hybrid composites in chloride medium. International Journal of Electrochemical Science. 2014;**9**:5663-5674

[25] Alaneme KK, Ademilua BO, Bodunrin MO. Mechanical properties and corrosion behaviour of aluminium hybrid composites reinforced with silicon carbide and bamboo leaf ash. Tribology in Industry. 2013;**35**(1):25-35

[26] Alaneme KK, Adewuyi EO. Mechanical behaviour of Al-Mg-Si matrix composites reinforced with alumina and bamboo leaf ash. Metallurgical and Materials Engineering. 2013;**19**(3):177-187

[27] Yogeshwaran S, Prabhu R, Natrayan L, R. Murugan mechanical properties of leaf ashes reinforced aluminum alloy metal matrix composites. International Journal of Applied Engineering Research. 2015;**10**(13):11048-11052

[28] Natrayan L, Sivaprakash V, Santhosh MS. Mechanical, microstructure and wear behavior of the material AA6061

reinforced SiC with different leaf ashes using advanced stir casting method. International Journal of Engineering and Advanced Technology (IJEAT). 2018; **8**(2S):366-371

[29] Poornesh M, Saldanha JX, Singh J, Pinto GM, Gaurav. Effect of coconut shell ash and SiC particles on mechanical properties of aluminium based composites. American Journal of Materials Science. 2017;**7**(4):112-115

[30] Nithyanandhan T, Ramamoorthi R, Sivasurya KV, Vesvanth M, Sanjay S, Sriram K. Machining behaviour of aluminium 6063 with strengthen aluminium oxide and chicken bone ash produced by modern technique. Journal of Science Technology and Research (JSTAR). 2021;**2**(1)

[31] Sulaiman A, Adeniyi AT, Toyin AA, Lanre S-BY, Kolawole AT, Idowu AI, et al. Hardness and tensile properties of prophylactic knee brace produced from cow bone and periwinkle shell composites. International Journal of Engineering Materials and Manufacture. 2019;**4**(2):41-47

[32] Arun P, Shanmughasundaram P, Suganthi S, Kumar R, Rogan VK. Wear Behaviour of Al-Chicken Bone Ash Metal Matrix Composites. Journal of Advances in Natural and Applied Sciences. 2016;**10** (6):11-15

[33] Atuanya CU, Ibhadode AOA, Dagwa IM. Effects of breadfruit seed hull ash on the microstructures and properties of Al–Si–Fe alloy/breadfruit seed hull ash particulate composites. Results in Physics. 2012;**2**:142-149

[34] Daramola OO, Adediran AA, Fadumiye AT. Evaluation of the mechanical properties and corrosion behaviour of coconut shell ash reinforced aluminium (6063) alloy composites. Leonardo Electronic Journal of Practices and Technologies. July – December 2015;(27):107-119

[35] Agunsoye JO, Talabi SI, Bello SA, Awe IO. The effects of Cocos Nucifera (coconut shell) on the mechanical and tribological properties of recycled waste aluminium can composites. Tribology in Industry. 2014;**36**(2):155-162

[36] Apasi A, Yawas DS, Abdulkareem S, Kolawole MY. Improving mechanical properties of aluminium alloy through addition of coconut shell-ash. Journal of Science and Technology. 2016;**36**(3):34-43

[37] Madakson PB, Yawas DS, Apasi A. Characterization of coconut Shell ash for potential utilization in metal matrix composites for automotive applications. International Journal of Engineering Science and Technology (IJEST). 2012; **4**(03):1190-1198

[38] Bello SA, Raheem IA, Raji NK. Study of tensile properties, fractography and morphology of aluminium (1xxx)/ coconut shell micro particle composites. Journal of King Saud University— Engineering Sciences. July 2017;**29**(3): 269-277. DOI: 10.1016/j. jksues.2015.10.001

[39] Mohanty H, Chowdhury S, Rout BK, Panda SK. Study on mechanical and machinability properties of aluminium coconut shell ash by Taguchi approach. International Journal of Innovative Research in Technology. 2017;**3**(11): 68-74

[40] Yekinni AA, Bello SK, Bajela GG, Adigun IA. Effect of coconut shell ash and graphite particles on microstructure and mechanical properties of recycled aluminium composites. International Journal of Advances in Scientific Research and Engineering (IJASRE). 2020;**6**(10):61-71

[41] Babaremu KO, Joseph OO, Akinlabi ET, Jen TC, Oladijo OP. Morphological investigation and mechanical behaviour of agrowaste reinforced aluminium alloy 8011 for service life improvement. Heliyon. 2020;**6**(e05506):1-8

[42] Fatile OB, Akinruli JI, Amori AA. Microstructure and mechanical behaviour of stir-cast Al-Mg-Sl alloy matrix hybrid composite reinforced with corn cob ash and silicon carbide. International Journal of Engineering and Technology Innovations. 2014;**4**(4):251-259

[43] Odoni BU, Odikpo EF, Chinasa NC, Akaluzia RO. Experimental analysis, predictive modelling and optimization of some physical and mechanical properties of aluminium 6063 alloy based composites reinforced with corn cob ash. Journal of Materials and Engineering Structures. 2020;7:451-465

[44] Hassan SB, Aigbodion VS. Effects of eggshell on the microstructures and properties of Al–Cu–Mg/eggshell particulate composites. Journal of King Saud University–Engineering Sciences. January 2015;**27**(1):49-56

[45] Dwiwedi S, Srivastava A, Sugimoto K, Chopkar M. Microstructural and mechanical characterization of chicken eggshell-reinforced Al6061 matrix composites. Open Journal of Metal. 2018; **8**:1-13

[46] Dwivedi SP, Sharma S, Mishra RK. Characterization of waste eggshells and $CaCO_3$ reinforced AA2014 green metal matrix composites: A green approach in the synthesis of composites. International Journal of Precision Engineering and Manufacturing;**17**: 1383-1393

[47] Dwivedi SP, Sharma S, Mishra RK. A comparative study of waste eggshells, CaCO3, and SiC-reinforced AA2014 green metal matrix composites. Journal of Composite Materials. 2017;**51**(17): 2407-2421. DOI: 10.1177/ 0021998316672295

[48] Dwivedi SP, Maurya NK, Maurya M. Effect of uncarbonized eggshell weight percentage on mechanical properties of composite material developed by electromagnetic stir casting technique. Revue des Composites et des Materiaux Avances;**29**(2):101-107

[49] Chaithanyasai A, Vakchore PR, Umasankar V. The micro structural and mechanical property study of effects of egg shell particles on the Aluminum 6061. Procedia Engineering. 2014;**97**: 961-967

[50] Alaneme KK, Fatile BO, Borode JO. Mechanical and corrosion behaviour of Zn-27Al based composites reinforced with groundnut Shell ash and silicon carbide. Tribology in Industry. 2014; **36**(2):195-203

[51] Singh J, Suri NM, Verma A. Affect of mechanical properties on groundnut shell ash reinforced Al6063. International Journal For Technological Research In Engineering. 2015;**2**(11): 2619-2623

[52] Alaneme KK, Eze HI, Bodunrin MO. Corrosion behaviour of groundnut shell ash and silicon carbide hybrid reinforced Al-Mg-Si alloy matrix composites in 3.5% NaCl and 0.3M H_2SO_4. Leonardo Electronic Journal of Practices and Technologies. January – June 2015;(26): 129-146

[53] Alaneme KK, Bodunrin MO, Adebimpe A. Awe, Microstructure, mechanical and fracture properties of groundnut shell ash and silicon carbide dispersion strengthened aluminium matrix composites. Journal of King Saud

University–Engineering Sciences. 2018; **30**:96-103

[54] Jose J, Christy TV, Eby Peter P, Feby JA, George AJ, Joseph J, et al. Manufacture and characterization of a novel agro-waste based low cost metal matrix composite (MMC) by compocasting. Materials Research Express. 2018;**5**(6):53-66

[55] Oghenevweta JE, Aigbodion VS, Nyior GB, Asuke F. Mechanical properties and microstructural analysis of Al–Si–Mg/carbonized maize stalk waste particulate composites. Journal of King Saud University—Engineering Sciences. 2016;**28**:222-229

[56] Ochuokpa EO, Yawas DS, Sumaila M, Adebisi AA. Development of aluminium based mango seed Mangiferaindica Shell ash (MSSA) particulate metal matrix composite. International Journal of Engineering Materials and Manufacture. 2021;**6**(3): 176-186

[57] Ochuokpa EO, Yawas DS, Okorie PU, Sumaila M. Evaluation of mechanical and metallurgical properties Al- Si-Mg/Mangiferaindica seed Shell ash (MSSA) particulate composite for production of motorcycle hub. Journal of Science Technology and Education. 2021;**9**(1):221-238

[58] Ochuokpa EO, Yawas DS, Samuel MSBO. Production and wear optimization of an MSSA-reinforced Al–Si–Mg composite: A Taguchi approach. The International Journal of Advanced Manufacturing Technology. 2022;**121**:4817-4828

[59] Salihu SA, Suleiman IY. Effect of Marula seed cake on the mechanical properties of aluminium alloys for the production of brake pad. Materials Science, International Journal of Scientific Research in Science, Engineering and Technology. 2017;**3**(6): 713-717

[60] Ramesh C, Valliappan M, Prasanna SC. Fabrication of ammcs by using stir casting method for hand lever. International Journal of New Technologies in Science and Engineering. 2015;**2**(1):1-6

[61] Edoziuno FO, Adediran AA, Odoni BU, Utu OG, Olayanju A. Physico-chemical and morphological evaluation of palm kernel shell particulate reinforced aluminium matrix composites. Materials Today: Proceedings. 2021;**38**:652-657

[62] Oladele IO, Okoro AM. The effect of palm kernel shell ash on the mechanical properties of as-cast aluminium alloy matrix composites. Leonardo Journal of Sciences. January – June 2016;(28):15-30. ISSN 1583-0233

[63] Aigbodion VS, Ezema IC. Multifunctional A356 alloy/PKSAnp composites: Microstructure and mechanical properties. Defence Technology. 2020;**16**:731-736

[64] Edoziuno FO, Odoni BU, Alo FI, Nwaeju CC. Dry sliding wear and surface morphological examination of an aluminium matrix composite reinforced with palm kernel shell. ACTA Metallurgica Slovaca. 2020;**26**(2):54-62

[65] Ajibola WA, Fakeye AB. Production and characterization of zinc-aluminium, silicon carbide reinforced with palm kernel shell ash. International Journal of Engineering Trends and Technology (IJETT). 2016;**41**(6):318-325

[66] Ebhojiaye RS, Amiolemhen PE, Ibhadode AOA. Development of hybrid composite material of palm kernel Shell (PKS) and periwinkle shell (PS) particles

in pure aluminium matrix. Solid State Phenomena. 2018;**278**:54-62 ISSN: 1662-9779

[67] Hemanth Kumar KS, Siddeswarappa B, Kulkarni PP. A review on mechanical behavior of aluminium 6061 reinforced with different agro ashes. International Journal for Research in Applied Science & Engineering Technology (IJRASET). 2020;**8**(XII):382-388. Available from: www.ijraset.com

[68] P. P. Student Ram Babu Matta Ravindra Babu Associate J A Ranga Babu, "Effect of Mechanical Properties on Palm Sprout Shell Ash Reinforced With Al-6061 Alloy Metal Matrix Composites", International Journal for Research in Applied Science and Engineering Technology (IJRASET), Volume 5, Issue 3, May – June 2019, 8–14

[69] Umunakwe R, Olaleye DJ, Oyetunji A, Okoye OC, Umunakwe IJ. Assessment of some mechanical properties and microstructure of particulate periwinkle shell-aluminium 6063 metal matrix composite (PPS-ALMMC) produced by two-step casting. Nigerian Journal of Technology (NIJOTECH). 2017;**36**(2): 421-427

[70] Umunakwe R, Olaleye DJ, Oyetunji A, Okoye OC, Umunakwe IJ. Assessment of some mechanical properties of particulate periwinkle shell-aluminium 6063 metal matrix composite (pps-almmc) produced by two-step casting. Nigerian Journal of Technology (NIJOTECH). 2017;**36**(2):421-427

[71] Yekinni AA, Durowoju MO, Agunsoye JO, Mudashiru LO, Animashaun LA, Sogunro OD. Automotive application of hybrid composites of aluminium alloy matrix: A review of Rice husk ash based reinforcements. International Journal of Composite Materials. 2019;**9**(2):44-52. DOI: 10.5923/j.cmaterials.20190902.03

[72] Nwabufor MN. Development and Characterization of AL−3.7% Cu−1.4% Mg Alloy/Periwinkle Ash (Turritella Communis) Particle Composites. Zaria: Ahmadu Bello University, the Department of Metallurgical and Materials Engineering; 2015

[73] Haque MH, Ahmed R, Khan MM, Shahriar S. Fabrication, reinforcement and characterization of metal matrix composites (MMCs) using rice husk ash and aluminium alloy. International Journal of Scientific & Engineering Research. 2016;**7**(3):28 ISSN 2229-5518 IJSER © 2016 Available from: http://www.ijser.org

[74] Siva Prasad D, Krishna AR. Production and mechanical properties of A356.2 /RHA composites. International Journal of Advanced Science and Technology. 2011;**33**:51

[75] Alanemea KK, Adewale TM. Influence of Rice husk ash—Silicon carbide weight ratios on the mechanical behaviour of Al-Mg-Si alloy matrix hybrid composites. Tribology in Industry. 2013;**35**(2):163-172 Available from: www.tribology.fink.rs

[76] Alaneme KK, Adewale TM, Olubambi PA. Corrosion and wear behaviour of Al–Mg–Si alloy matrix hybrid composites reinforced with rice husk ash and silicon carbide. The Journal of Materials Research & Technology. 2014;**3**(1):9-16

[77] Alanemea KK, Olubambi PA. Corrosion and wear behaviour of rice husk ash—Alumina reinforced Al–Mg–Si alloy matrix hybrid composites. The Journal of Materials Research & Technology. 2013;**2**(2):188-194

[78] Adekunle Y, Taiwo R, Ismail A, David S, Rasheed S, Olalekan A. Recycled aluminium cans/rice husk ash: evaluation of physico-mechanical properties. World Journal of Engineering Research and Technology. 2019;**5**(1):18-32

[79] Sarkar S, Bhirangi A, Mathew J, Oyyaravelu R, Kuppan P, Balan ASS. Fabrication characteristics and mechanical behavior of Rice Husk Ash-Silicon Carbide reinforced Al-6061 alloy matrix hybrid composite, Materials Today Proceedings. Part 2. 2018;**5**(5): 12706 -12718

[80] Chinta ND, Prasad KS, Kumar VM. Characterization of rice husk ash and SiC reinforced aluminium metal matrix hybrid composite. SSRG International Journal of Mechanical Engineering (SSRG-IJME). 2017;(Special Issue May): 329-332

[81] Usman AM, Raji A, Waziri NH, Hassan MA. Aluminium alloy—Rice husk ash composites production and analysis. Leonardo Electronic Journal of Practices and Technologies. July – December 2014;(25):84-98

[82] Subrahmanyam APSVR, Madhukiran J, Naresh G, Madhusudhan S. Fabrication and characterization of Al356.2, rice husk ash and fly ash reinforced hybrid metal matrix composite. International Journal of Advanced Science and Technology. 2016;**94**:49-56. DOI: 10.14257/ijast.2016.94.05

[83] Saravanana SD, Senthil Kumar M. Effect of mechanical properties on Rice husk ash reinforced Aluminum alloy (AlSi10Mg) matrix composites. Procedia Engineering. 2013;**64**:1505-1513. DOI: 10.1016/j.proeng.2013.09.232

[84] Senapati AK, Sahoo SK, Singh S, Sah S, Padhi PR, Satapathy N. A comparative investigation on physical and mechanical properties of MMC reinforced with waste materials. International Journal of Engineering and Advanced Technology (IJEAT). 2017;**6**(4) ISSN: 2249-8958

[85] Asafa TB, Durowoju MO, Oyewole AA, Solomon SO, Adegoke RM, Aremu OJ. Potentials of Snailshell as a reinforcement for discarded Aluminum based materials. International Journal of Advanced Science and Technology. 2015; **84**:1-8. DOI: 10.14257/ijast.2015.84.01

[86] Ahmad Faud MY, Yaakob I, Mohd Ishak ZA, Mohd Omar AK. Density measurement of rice husk ash filler particles polypropylene composites. Polymer Testing. 1993;**12**:107-112

[87] Umoh AA, Jonah UM. Use of coconut fiber as reinforcement in bamboo leaf ash blended cement- based composite panels. Jordan Journal of Civil Engineering. 2015;**9**(4):435-444

[88] Alaneme K, Babajide A, Omotoyinbo A, Borode JO, Bodunrin M. Corrosion and wear behavior of stir-cast aluminium-based hybrid composites reinforced with silicon carbide and corn cob ash. African Corrosion Journal. 2015; **1**(2):1–8

[89] https://www.sciencedirect.com/topics/engineering/corn-cob

[90] Krishnamoorthy B, Ramasamy V. Feasibility study on eggshell powder concrete (ESP concrete). IJSRD—International Journal for Scientific Research & Development. 2019;**6**(12): 732-734

[91] Kumar TVA, Aradwad PP, Thirupathi V, Rajkumar P, Viswanathan R. Briquetting of mango seed shell: Effect of temperature, pressure and binder. Agricultural Engineering Today, ISAE. 2021;**45**(4):23-28

[92] https://www.aqua-calc.com/page/density-table/substance/peanut-blank-shells

[93] Oyejobi DO, Abdulkadir TS, Yusuf IT, Badiru MJ. Effects of palm kernel shells sizes and mix ratios on lightweight concrete. Journal of Research Information in Civil Engineering. 2012; **9**(2):217-226

[94] Owoyemi HT, Owoyemi AG. Chemical and phase characterization of snail shell (Archachatina Marginata) as bio-waste from South-West in Nigeria for industrial applications. Chemistry and Materials Research. 2020;**12**(6):15-20

[95] Walton R. The impact of aluminium and silicon on the shell strength of *Lymnaea stagnalis*. The Malacologist, The Malacological Society of London. 2009; **53**. Available from: http://malacsoc.org.uk/the_Malacologist/BULL53/walton.htm

[96] Anon. A note on the density of bovine limb bones. Animal Science. 1975; **21**(2):195-198. DOI: 10.1017/S0003356100030622

[97] Perkarskas J, Zibutis S, Grazuleviciene V, Grigalavviciene I, Mazeika R. Cattle horn shavings as slow release nitrogen fertilizer. Polish Journal of Environmental Studies;**24**(2):645-650

[98] Mason P. Density and structure of alpha-keratin. Nature. 1963;**197**:179-180

[99] Alaneme KK, Bodunrin MO. Mechanical behaviour of alumina reinforced AA6063 metal matrix composites developed by two-step stir casting process. ACTA Bulletin of Engineering, Tome VI. (July – September) 2013:105-110

[100] Alaneme KK, Sanusi KO. Microstructural characteristics, mechanical and wear behaviour of aluminium matrix hybrid composites reinforced with alumina, rice husk ash and graphite. Engineering Science and Technology, An International Journal. 2015;**xxx**:1-7

[101] Ajagol P, Anjan BN, Rajaneesh NM, Preetham KGV. Effect of SiC reinforcement on microstructure and mechanical properties of Aluminum metal matrix composite. International Conference on Advances in Manufacturing, Materials and Energy Engineering (ICon MMEE 2018) 2–3 March 2018, MANGALORE INSTITUTE OF TECHNOLOGY & ENGINEERING, Badaga Mijar, Moodbidri, Karnataka, India). IOP Conference Series: Materials Science and Engineering. 2018;**376**:1-9

[102] Yung CK, Chan SLI. Tensile properties of nanometric Al2O3 particulate reinforced aluminium matrix composites. Journal of Materials Chemistry and Physics. 2004;**85**:438443

[103] Qu S, Geng L, Han J. SiC/Al composite fabricated by modified squeeze casting technique. Journal of Materials Science and Technology. 2007; **23**(5):641-644

[104] Rohatgi P, Asthana R, Yarandi E. Solidification During Casting of Metal-Matrix Composites. American Society of Metals (ASM) Handbook. Volume 15: Casting ASM Handbook Committee. ASM International. 2008:390-397

[105] Anon. Internal Combustion Engine Parts. Upload. Lok Nayak Jai Prakash Institute of Technology (LNJPIT), Chapra, Bihar, India. Available from: https://www.lnjpitchapra.in/wp-content/uploads/2020/05/file_5eb3d452ac7a3.pdf. [Accessed: 23 July 2022]

[106] London Metal Exchange (LME) Alumina (Platts). Available from: https://www.lme.com/Metals/Non-

ferrous/LME-Alumina#Trading+day +summary. [Accessed: 25 July 2022]

[107] Anon. Global Silicon Carbide Markets, 2021-2027 - Growing Steel Industry & Growing Demand from Semiconductor and Electronics Industry, Yahoo Finance. Available from: https://finance.yahoo.com/news/global-silicon-carbide-markets-2021-

Chapter 6

Optimal Integration of Series and Shunt FACTS with Wind Energy for Active Power Loss Reduction

I. Made Wartana and Ni Putu Agustini

Abstract

The integration of wind energy (WE) with flexible AC transmission system (FACTS) devices into the grid to improve grid performance is one of the latest advances in renewable energy (RE) technology. This work proposes the optimal placement and size of a WE, a doubly-fed induction generator (DFIG) with two of FACTS controller, viz. thyristor-controlled series compensator (TCSC) static var compensator (SVC). The goal is to maximize system bus load (Max. LBS) and minimize active power loss (Min. P_{loss}) by satisfying various safety and stability constraints. Newton Raphson's power flow study involving TCSC, SVC, and DFIG is a bi-objective that meets multiple constraints: lines, generation, voltage limits, and small signal stability. A variant of the genetic algorithm, non-dominated sorting GA II (NSGA-II), was applied to solve the contradictory bi-objective optimization problem. A modified standard and practical test system, the IEEE 14-bus and the Indonesia Java Bali 24-bus, integrated with DFIG, TCSC, and SVC, were simulated to investigate the efficacy of the suggested technique. The simulation shows that the optimal placement and size of DFIG with both FACTS can improve system performance with all system loading conditions and meet all system constraints.

Keywords: doubly-fed induction generator, load bus system, small signal stability, wind energy, unified power flow controller

1. Introduction

In the last decade, recent advances in renewable energy (RE) technology have discussed the development of the technology industry, which includes conversion, storage, production, and management, especially wind energy systems, solar, hydropower, geothermal energy, bioenergy, and hydrogen production [1, 2]. The development of this large-scale energy technology and its impact on the global economy and electricity capacity is a fascinating study, especially related to system optimization and sizing, resource assessment and deployment, instrumentation and control, modeling and simulation, regulation, and policy [3–5].

The development of renewable energy sources (RES) resources into the mainstream electricity sector has increased in recent years as awareness of environmental issues and efforts to reduce dependence on fossil fuel resources increase. Wind energy (WE) is assumed to have the most commercial technics and economic prospects among the various renewable resources. Integrating the RES represented as a Distributed Generation (DG) into the grid has received wide attention and scope in power systems [6, 7]. In addition, DG can help utilize distributed energy sources with relatively small resources and reduce the use of electrical power transmission capacity [8]. Most DGs are located near the load center, and some types of DGs also provide reactive power to support the power system [9].

On the other hand, since DGs are installed close to the load, they can minimize active power loss (Min. P_{loss}) and increase profile voltage, improving system performance [10]. Another advantage of integrating DGs into the grid is that they can delay investments in transmission lines and the construction of large power plants. Above all, the second option/advantage has been utilized in this chapter to maximize the load bus system (Max.-LBS) by optimal placement of DGs into the grid [11]. However, some DGs, particularly wind generation systems, do not generate almost negligible reactive power. Numerous other mechanisms have been used to compensate the system when the DG can no longer support the system's reactive power requirements.

The integration of various DG technologies and flexible alternating current transmission systems (FACTS) into the grid started one of the significant management concerns for professional engineers [12, 13]. Various studies have extensively studied grids involving RES and FACTS devices to improve security, power quality, system stability, and optimal reactive power delivery using different latest optimization techniques [9, 10]. Some essential technical benefits are increasing the voltage profile with Min. P_{loss} and Max.-LBS. All the while enhancing the power system's security, reliability, and stability, thereby increasing the overall energy efficiency [14], furthermore relieving transmission and distribution congestion. In addition, DG utilizes the small energy resources available near the load center, thereby increasing energy access [15, 16]. Moreover, by installing FACTS series and shunts in the system, apart from increasing the performance of the system, it can also help secure and restore network operations during emergencies and after power outages. However, integrating a wide variety of DG technology and FACTS into the grid has become one of the significant management concerns for operational people.

Numerous works have been done on the optimal allocation of DG for different purposes. Different approach techniques have been suggested, that is, Genetic Algorithm (GA) [17], PSO [18], and an efficient hybrid approach distribution networks [19] for optimal location and settings of multi-types of DG. The optimal DG placement has been compared using CSA, GSA, PSO, and the firefly algorithm for the minimum real power loss in a radial distribution system [20, 21]. However, the system stability and security constraints have not been exclusively considered by Max.-LBS within any grid condition and their impact on the transmission loss with the DG placement.

From these literary works, it can be observed that most of the problems in optimal DG locations are often disclosed separately as a matter of mono-objective optimization [22–24]. Awkwardly, formulating the problems as a mono-objective optimization is not entirely practical. However, it is always good to utilize optimal DG placement with two conflicting objectives, considering the security and stability of the system formulated as a multi-objective problem and solved simultaneously [10].

In this work, a multi-objective problem has been formulated by Max. LBS by optimal location and settings of a DG, viz. wind generation system or farm, while maintaining the system security and stability margin within an acceptable range. Utilizing DG optimal placement, the Max. LBS and the Min. P_{loss} can be done simultaneously. The multi-objective problems have been solved simultaneously using the novel variant of GA specialized in multi-objective optimization, namely the NSGA-II by optimal location and sizing of the DG.

2. System component modeling

2.1 Wind turbine generator modeling

Of the two basic types of wind turbine generators (WTG) available, the second category is variable speed. This popular type of WTG is known as a double induction generator (DFIG) and synchronous direct-drive generator [25]. This type in addition to offering increased efficiency in capturing energy from the wind over a broader range of wind speeds also has better power quality and the ability to regulate power factors, both consuming and generating reactive power. **Figure 1** shows a DFIG type of WTG, where the rotor through a back-to-back ac/dc/ac converter is connected to the grid, while the stator is connected directly to the grid [26, 27]. In this model, the generator is decoupled from the grid since the dynamics of the stator, and rotor flux is fast compared to the grid dynamics and the converter controls; consequently, the steady-state electrical equation of the DFIG is assumed as (1):

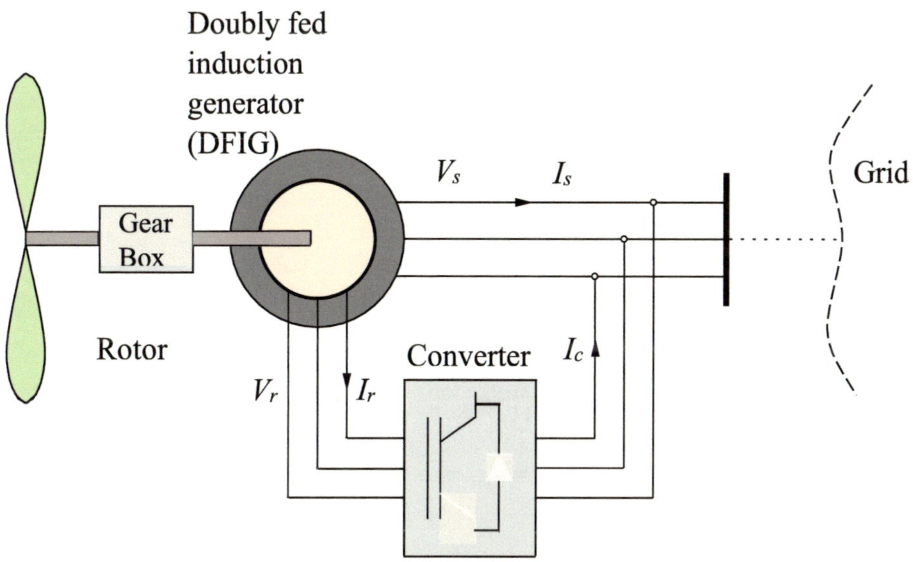

Figure 1.
Variable speed of WTD with DFIG modeling.

$$\left.\begin{aligned} v_{ds} &= -R_s i_{ds} + \left((X_s + X_m) i_{qs} + X_m i_{qr}\right) \\ v_{qs} &= -R_s i_{qs} + \left((X_s + X_m) i_{ds} + X_m i_{dr}\right) \\ v_{dr} &= -R_R i_{dr} + (1 - \omega_m)\left((X_R + X_m) i_{qr} + X_m i_{qs}\right) \\ v_{qr} &= -R_R i_{qr} + (1 - \omega_m)\left((X_R + X_m) i_{dr} + X_m i_{ds}\right) \end{aligned}\right\} \tag{1}$$

where v_{ds} and v_{qs} are d-axis and q-axis stator voltage, respectively, v_{dx} and v_{qr} are d- and q-axis rotor voltage, respectively, i_{dx} and vi_{qx} are d-axis and q-axis rotor current, respectively, R_s and R_r are stator and rotor resistance, respectively, X_s and X_m are stator and magnetization reactance, respectively, X_r and ω_m are rotor reactance and the rotor speed, respectively.

Whereas both v_{ds} and v_{qs} are functions of the grid magnitude and phase as formulated in Eq. (2):

$$\left.\begin{aligned} V_{ds} &= V \sin(-\theta) \\ V_{qs} &= V \cos(-\theta) \end{aligned}\right\} \tag{2}$$

The generator's active and reactive power depends on the stator and current converter, as shown in Eq. (3):

$$\left.\begin{aligned} P &= v\,i + v\,i + v\,i + v\,i \\ Q &= v_{qs} i_{ds} + v_{ds} i_{qs} + v_{qc} i_{dc} + v_{dc} i_{qc} \end{aligned}\right\} \tag{3}$$

2.2 Thyristor controlled series compensator (TCSC) modeling

One popular type of FACTS series will be used in this study, namely the thyristor-controlled series compensator (TCSC). The function of installing these devices into the grid, not only to increase the series capacitance, but also to reduce the total reactance of the grid. Moreover, TCSC is also used for achieving maximum flexibility in managing the grid reactance [28]. Although the cost of the device was a little bit more expensive, compare similar the FACTS, they provide real-time capabilities to respond to the grid conditions. It is done by dynamically adjusting the amount of reactance compensated [29]. **Figure 2** shows TCSC modeling as a constant capacitive reactance on a transmission line connected between bus-i and j. This capacitive reactance modifies the line reactance x_{ij}, during power flow analysis, as stated in (4) [30].

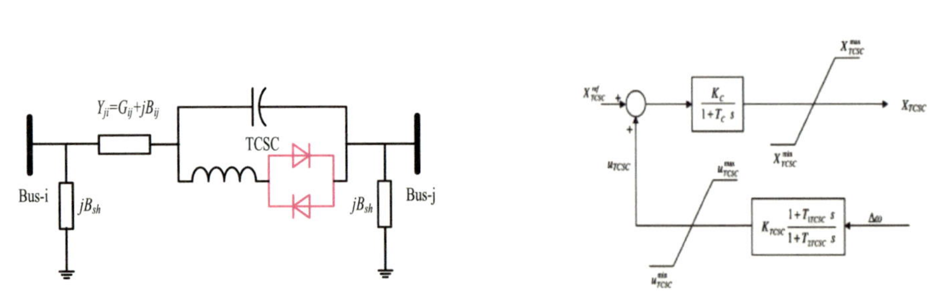

(a) (b)

Figure 2.
TCSC modeling. (a) TCSC model in power system; (b) block diagram model of TCSC.

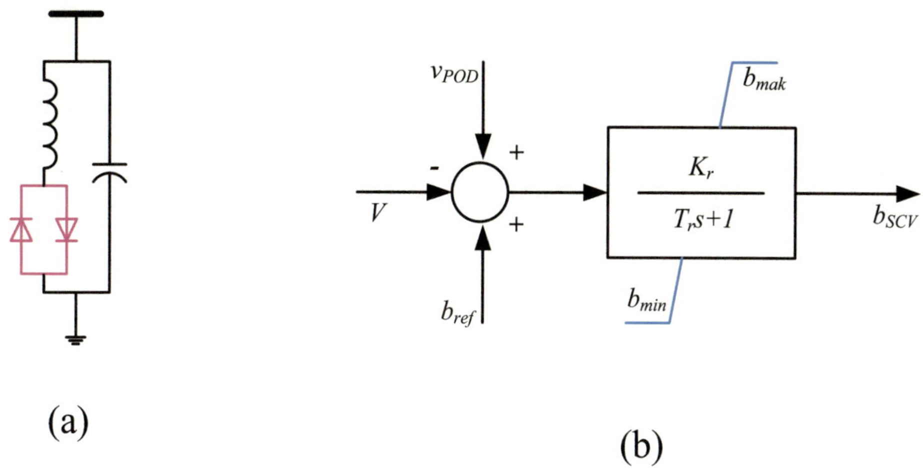

Figure 3.
SVC modeling. (a) SVC model in power system; (b) block diagram model of SVC.

$$x'_{ij} = (1 - c_p)x_{ij} \tag{4}$$

2.3 Static var compensator (SVC) modeling

SVC is also a popular type of FACTS which is a shunt-connected variable reactor to the bus that can inject or absorb reactive power to regulate the bus voltage. Because it can provide instant reactive power for voltage support having two regions, capacitive and inductive, either in capacitive or inductive mode, SVC can inject reactive and inductive power [31]. **Figure 3** depicted the SVC modeling as an equivalent series of susceptance variables depending on the node requirements. The differential and algebraic equations formulated in (5) and (6) in the model give the total reactance b_{SVC} and reactive power injected at the SVC node [32].

$$\dot{b}_{SVC} = \left(K_r\left(V_{ref} + v_{POD} - V\right) - b_{SVC}\right)/T_r \tag{5}$$

$$Q = b_{SVC}V^2 \tag{6}$$

where the maximum and minimum susceptance (p.u) are presented by b_{max} and b_{min}, respectively, and the input signal for power system oscillation damping is modeled as v_{POD}.

3. Problem formulation

Increasing the load to attain the maximum load bus system (Max. LBS) but with minimum line power loss (Min. P_{loss}) was preferred as two conflicting optimization objectives. To solve the optimization algorithm developed whereas maintaining the security and stability of all systems, NSGA-II can handle it simultaneously. The problem is formulated as a discrete-continuous multi-objective optimization with the real constraint $F(\mathbf{x}, \mathbf{u})$. While the dependent and control variables are represented by $\mathbf{x}$ and $\mathbf{u}$ as formulated in Eqs. (7) and (8), respectively [33].

$$\textit{Minimize}\ F(\mathbf{x}, \mathbf{u}) = [-F_1(\mathbf{x}, \mathbf{u}), F_2(\mathbf{x}, \mathbf{u})] \tag{7}$$

$$\textit{Subject to}: g_i(\mathbf{x}, \mathbf{u}) = 0\ \mathrm{i} = 1, \dots, \mathrm{M}$$

$$h_j(\mathbf{x}, \mathbf{u}) \le 0\ \mathrm{j} = 1, \dots, \mathrm{N} \tag{8}$$

where the objective functions to be optimized are presented by F_1 and F_2, respectively. The ith equality and the jth inequality constraints are gi and h_j, respectively. The number of equality and inequality constraints are M and N, respectively.

3.1 Max. LBS

This research is to maximize the load bus system (Max. LBS) chosen as the first objective function with a load increase scenario as formulated in (9) and (10), respectively:

$$\textit{Maximize}\ F_1(\mathbf{x}, \mathbf{u}, \lambda) \tag{9}$$

$$\textit{Subject to}: VL = \sum_{i=1}^{N_L} OLL_i + \sum_{i=1}^{N_L} BVV_i \tag{10}$$

where the system load parameter is λ, derived in (9), VL, which are the sum of OLL_i and BVV_j stated at (12) and (13), respectively. Both characterize the thermal and bus violation limit factors. The total number of transmission lines and load busses are indicated as N_L and N_E, respectively [34].

$$\lambda = \exp\lfloor\gamma\left|\lambda_f - \lambda_f^{\max}\right|\rfloor; \lambda_f \in \left[1, \lambda_f^{\max}\right] \tag{11}$$

where the slope adjustment coefficient of a function is denoted as γ, the active and reactive power demand, P_{Di} and Q_{Di}, as formulated in (12) and (13), respectively revealed in the load factor λ_f which has a maximum value of λ_f^{max}.

$$P_{Di}(\lambda_f) = (\lambda_f)P_{Di} \tag{12}$$

$$Q_{Di}(\lambda_f) = (\lambda_f)Q_{Di} \tag{13}$$

The first term of Eq. (10), OLL_i, formulated at Eq. (13) represents the indices of the system security state in which its value is equal to 1 if the jth line loading is less than its rating. Otherwise, it increases logarithm with the overload as given in (14):

$$OLL_i = \begin{cases} 1;\text{if } P_{ij} \le P_{ij}^{\max} \\ \exp\left(\Gamma_{OLL}\left|1 - \frac{P_{ij}}{P_{ij}^{\max}}\right|\right);\text{if } P_{ij} \ge P_{ij}^{\max} \end{cases} \tag{14}$$

where P_{ij} and P_{ij}^{max} are the real power flow between bus-i and j and its thermal limit, respectively. Whereas the coefficient used to adjust the slope of the exponential function is Γ_{OLL}. The second term, BVV_j, state of (10), is the indices of the system security associated with the bus voltage violation factor at bus-j as defined in (15):

$$BVV_i = \begin{cases} 1;\text{if } 0.9 \le V_b \le 1.1 \\ \exp(\Gamma_{BVV}|1 - V_b|);\text{otherwise} \end{cases} \tag{15}$$

where Γ_{BVV} is the coefficient used to adjust the slope of the exponential function, similar to Eq. (13), If BVV_j is equal to 1, the voltage level drops between their minimum and maximum limits; otherwise, the voltage deviation increases exponentially.

3.2 Min. P_{loss}

The second objective function is to minimize the line power loss (P_{loss}) of the transmission line as expressed in (16) [35].

$$F_2(\mathbf{x}, \mathrm{u}) = P_{loss} = \sum_{k=1}^{nl} g_k \left[V_i^2 + V_j^2 - 2V_iV_j \cos\left(\delta_1 - \delta_j\right)\right] \tag{16}$$

where nl is the transmission line numbers, g_k is the conductance of kth line; whereas $V_i\angle\delta_i$ is the voltage on the end bus-i and $V_j\angle\delta_j$ is the voltage on the end bus-j of the kth line.

3.3 The equality constraint

At every node, the typical load flow eqs. [36] are denoted as their equality constraints expressed in (17).

$$\begin{aligned} P_{Gi} &= P_{Li} + V_i \sum_{i=1}^{Nb} V_j\left(G_{ij}\cos\delta_{ij} + G_{ij}\sin\delta_{ij}\right); \quad i = 1,2,\cdots N_b \\ Q_{Gi} &= Q_{Li} + V_i \sum_{i=1}^{Nb} V_j\left(G_{ij}\cos\delta_{ij} + G_{ij}\sin\delta_{ij}\right); \quad i = 1,2,\cdots N_b \end{aligned} \tag{17}$$

where N_b is the of busses number.

3.4 Small-signal stability

This stability constraint represents the dynamic performance of the power system, and is related to the weakening of the electromechanical oscillation mode. The oscillation behavior is similar to (i) the variation of electrical torque developed by the synchronous machines when the rotors angle changes; and (ii) the inertia of the rotor. This oscillation mode is usually associated with a range of frequencies between 0.5 and 4 Hz [37]. A set of differential-algebraic equations (DAE), used for small signal stability analysis, is formulated as (1):

$$\left.\begin{aligned} x &= f(x, y) \\ 0 &= g(x, y) \end{aligned}\right\} \tag{18}$$

where the vectors of the state and algebraic variables are represented by x and y, respectively. To calculate the state matrix A_s uses the complete Jacobian matrix manipulation A_C by determining the linearization of the DAE system in (19):

$$\begin{bmatrix} \Delta x \\ 0 \end{bmatrix} = \begin{bmatrix} \nabla_x f & \nabla_y f \\ \nabla_x g & \nabla_y g \end{bmatrix} \begin{bmatrix} \Delta x \\ \Delta y \end{bmatrix} = [A_c] \begin{bmatrix} \Delta x \\ \Delta y \end{bmatrix} \tag{19}$$

By simply eliminating the algebraic variables of the state matrix A_s the (20) attained. Implicitly this equation assumes that the Jacobian model's power flow is non-singular:

$$A_s = F_x - F_y G_y^{-1} G_x \tag{20}$$

The matrix leads to the calculation of eigenvalues in the S-domain, which states that if the real part of the eigenvalues is less than 0, then the system is stable.

3.5 Fast voltage stability index

Fast Voltage Stability Index (FVSI) [38], is one of the stability indices used to guarantee safe bus loading in this study as formulated in (21):

$$FVSI_{ij} = \frac{4Z^2 Q_J}{V_I^2 X} \tag{21}$$

Since its value is close to 1.00 specifies that the line is potential the point of instability, but if the total goes above 1.00, a quick voltage drop can happen on one of the busses coupled to the line producing the system to collapse.

3.6 Line stability factor (LQP)

Eq. (22) formulates the line stability factor (LQP) applied to [39], to ensure the system stability index if the value is less than 1.00:

$$LQP_{ij} = 4\left(\frac{X}{V_i^2}\right)\left(\frac{X}{V_i^2}P_I^2 + Q_j\right) \tag{22}$$

4. Brief description of the NSGA-II

The technique developed by K. Deb [40] is included in the Elitist multi-objective evolutionary algorithm category with a fast non-dominated sorting approach with low computational complexity. In the NSGA-II, elite preservative operators favor the elite of a population by allowing them to be taken directly to the next generation. In this process,good solutions that are found in the beginning will never disappear unless better solutions are found. The bi-objective of the multi-objective optimization algorithm is maintained by using a fitness assignment scheme that prefers a non-dominated solution. The population in this technique is initialized as usual as an evolution algorithm solves multi-objective optimization problems [41]. The NSGA-II procedure is presented in **Figure 4**.

For every generation k does:

a. With size N, create a random parent population;

b. Based on non-dominance, then sort the population;

c. With its non-dominant level (assumed to minimize fitness), assign each fitness solution (or rating) the same;

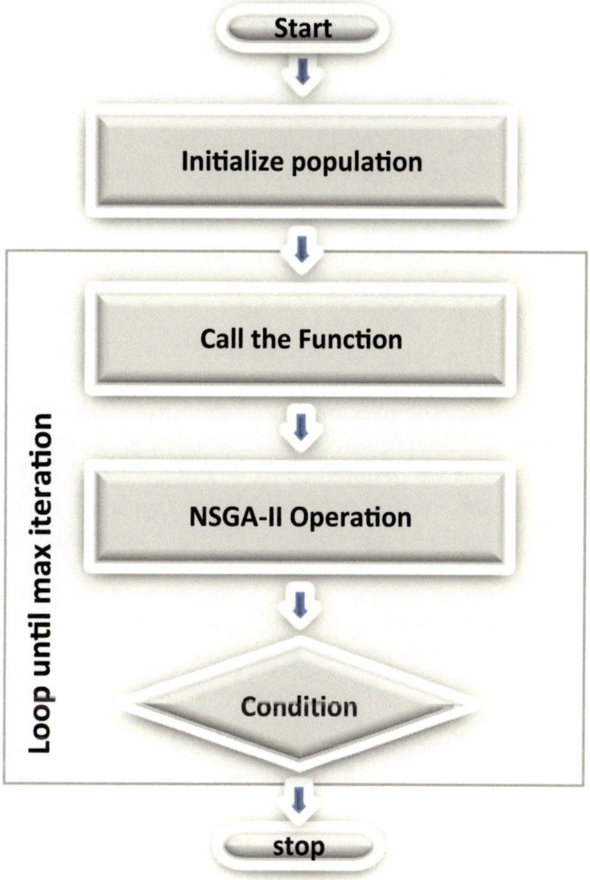

Figure 4.
NSGA-II procedure.

d. Create a new offspring population of size N using the usual binary tournament selection, recombination, and mutation operators for;

e. To form an expanded population of size 2 N, combine the offspring and parent populations;

f. Based on non-dominance, sort the expanded population;

g. With the individuals from the sorting sequence starting from the best, fill in the new population of size N;

h. To ensure diversity, if the front can only partially fill the next generation (This strategy is called "niching"), call the crowding comparison operator;

i. So that until the stopping criteria are met, perhaps a certain number of generations, repeat steps (2) to (8).

Determining a compromise solution suitable for all non-inferior alternatives depends on the decision-making agency's subjective preferences, the decision maker (DM), so it is not just a matter of hanging. It is reasonable to assume that DM may have imprecise or objective ambiguity for each objective function due to the ambiguous nature of DM assessment. Thus, each membership function is not only determined by intuitive knowledge but also by the experience of DM, which is introduced to represent the purpose of each objective function, as formulated in (23) [42].

$$\mu_i = \begin{cases} 1 & F_i \le F_i^{\min} \\ \dfrac{F_i^{\max} - F_i}{F_i^{\max} - F_i^{\min}} & F_i^{\max} < F_i < F_i^{\min} \; in \\ 0 & F_i \ge F_i^{\max} \end{cases} \tag{23}$$

where, F_i^{max} and F_i^{min} are the maximum and minimum values of the ith objective function among all non-dominated solutions, respectively. A membership function that diverges between 0 and 1 is denoted by μ_i, where $\mu = 1$ and $\mu = 0$ signify complete compatibility and mismatch, respectively, of solutions with the set [43]. Meanwhile, Eq. (24) formulates a normalized membership function k for each calculated non-dominated solution k.

$$\mu^k = \frac{\sum_{i=1}^{N_{obj}} \mu_i^k}{\sum_{k=1}^{M} \sum_{i=1}^{N_{obj}} \mu_i^k} \tag{24}$$

where, the amounts of objective functions and non-dominated solutions are denoted as N_{obj} and M, respectively. A membership function of non-dominated solutions in a fuzzy set can be deliberated in the function μ^k, where the solution devouring the maximum membership in the fuzzy set is deliberated as the best compromise solution.

a. The NSGA-II implementation for optimal integration of WTG with FACTS.

Such as validated in this study, the bi-objectives of this study are to attain Max. LBS by optimal location and settings of one type of WTG with two series or shunt FACTS controller into the grid to minimize the P_{loss} (Min. P_{loss}) of transmission lines while keeping the system security and stability and margins. Eqs. (7) and (8) previously formulated the optimization problems which are expressed as a bi-objective optimization problems. To investigate both conflicting objectives simultaneously, the NSGA-II techniques have been employed in numerous case studies: i) base case: without WTG and FACTS device; ii) Case-1: WTG only; iii); Case-2: TCSC only; Case-3: WTG and TCSC, Case-4: SVC only and Case-5: WTG and SVC. The point of best compromise solution (best CS) from the Pareto front is associated with the optimal solution of the equivalent objectives optimized with Max. LBS is the best LBS, and Min. P_{loss} is the best P_{loss} for respectively of the bi-objectives.

5. System studies and result discussions

The technique proposed in this study has been tested on the modification of the standard test system IEEE 14-bus [36] and the practical test system, namely the

Indonesia Java-Bali 24-bus system [22]. The loads are denoted as constant power or PQ loads with constant power factors and changed based on Eqs. (11) and (12). Furthermore, one type of wind field, DFIG, can transmit large amounts of active power to the system and consumes and produces reactive power optimally integrated into the grid. Simultaneously two types of series and shunt of FACTS devices, TCSC or SVC, are integrated into the grid to control various stability and security of the system due to increased LBS. The decision variables are the location, sizing, and settings of DFIG and the FACTS. The number of the FACTS is fixed at one, and their constraints are chosen at the beginning. All the load busses system is nominated to be the optimal location of DFIG or SVC placement. At the same time, all the grid lines are nominated to be the optimal location of the TCSC controller, which is deliberated as a discreet variable. The parameters of NSGA-II for entirely optimization cases are given in **Table 1**.

5.1 IEEE 14-bus system

5.1.1 Base case: Without FACTS devices and WTG

In the base case condition, where the grid is not connected to FACTS devices and WTG, Pareto fronts are gained from the simulation of the NSGA-II technique, as depicted in **Figure 5**. Max. LBS and Min. P_{loss} was obtained at 149.59% and 0.1704%, respectively, but the best CS achieved by P_{loss} was slightly lower than the previous result of 1.1625%, as well as the LBS which slightly decreased by 111.51%. All system stability in this condition is not considered.

5.1.2 Case-1: WTG only

Connecting the WTG type DFIG on the grid obtained a non-dominated solution, as shown by the Pareto front in **Figure 6**. In this case, the best LBS and P_{loss} gained is 157.40% and 0.2239% by placing WTG on busses 14 and 8, respectively. The best value is similar to the previous base case; the increase in LBS is much higher even though the value of the P_{loss} has also increased slightly. Based on **Figure 6**, it can also be seen that the best CS value for P_{loss} is almost similar to the base case value of 0.1772%. The LBS value is slightly lower at 112.24% with optimal WTG location on bus 8 with both active and reactive power of 49.91 MW and −11.46 MVar, respectively. **Figure 7** proves that in this condition, the system is stable in terms of small-signal stability, which is expressed by negative values of all pairs of eigenvalues in the S-plane.

5.1.3 Case-2: TCSC only

Optimal placement of TCSC, as one type of FACTS series, on the grid for Case-2 has been carried out and produces a Pareto front as given in **Figure 8**. From the figure,

Population	Generations	Pool size	Tour size	η_c
100	50	25	2	20

Table 1.
NSGA-II parameters.

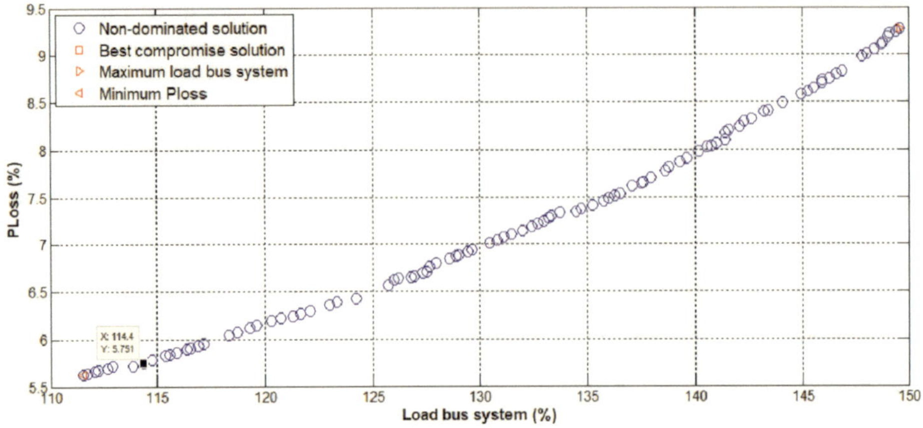

Figure 5.
Pareto front of the base case.

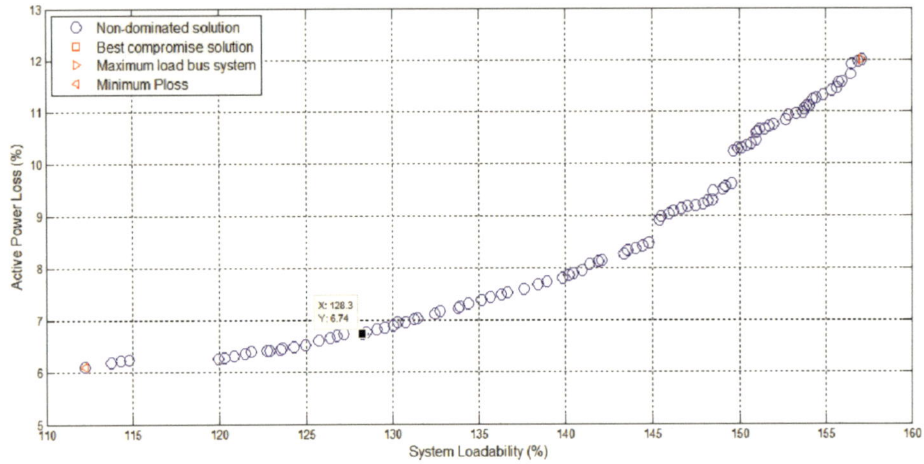

Figure 6.
Pareto front of Case-1.

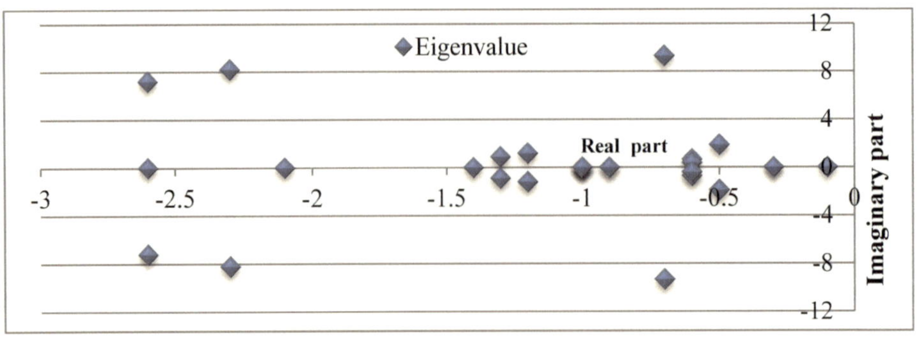

Figure 7.
The eigenvalue of Case-1.

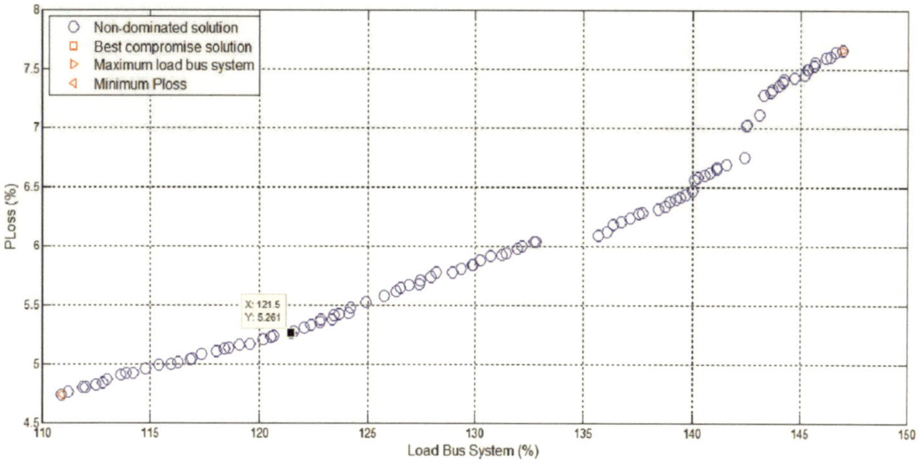

Figure 8.
Pareto front of Case-2.

it can be seen that for the best CS, optimal placement TCSC of 19.9599 p.u. on lines 15–12 results in an increase in LBS and P_{loss} of 121.49% and 0.1655%, respectively. The P_{loss} is better compared to the same solution in both previous cases; even though the LBS are slightly lower than in Case-1. While **Figure 9** depicts the eigenvalues for all conditions that meet the constraints of small-signal stability.

5.1.4 Case-3: WTG and TCSC

In case 3, with an optimal WTG placement of 54.55 MW and −16.06 MVar on bus-4 and a TCSC on lines 2–3 with an optimal setting of 20 p.u. is obtained the best CS for LBS and P_{loss} of 130.9% and 6.805%, respectively, as illustrated in **Figure 10**. The LBS value achieved is the highest value of all cases for the IEEE 14-bus test system, although the Ploss is almost similar to the results in case 1. **Figures 11–13** present numerous system stability constraints, not only eigenvalues but also FVSI and LQP of less than one, which indicate the system is stable under different conditions for the best CS.

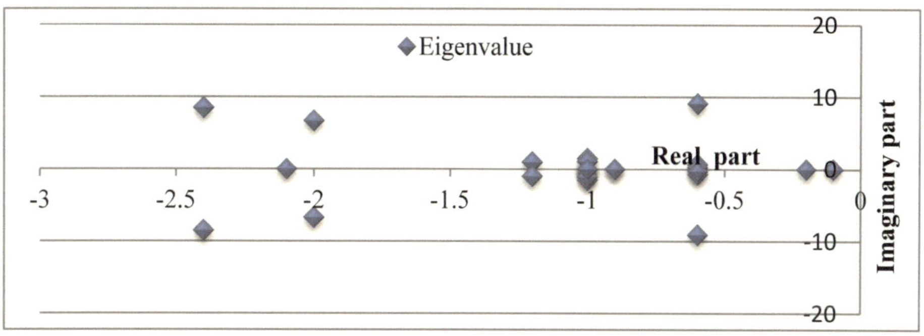

Figure 9.
The eigenvalue of Case-2.

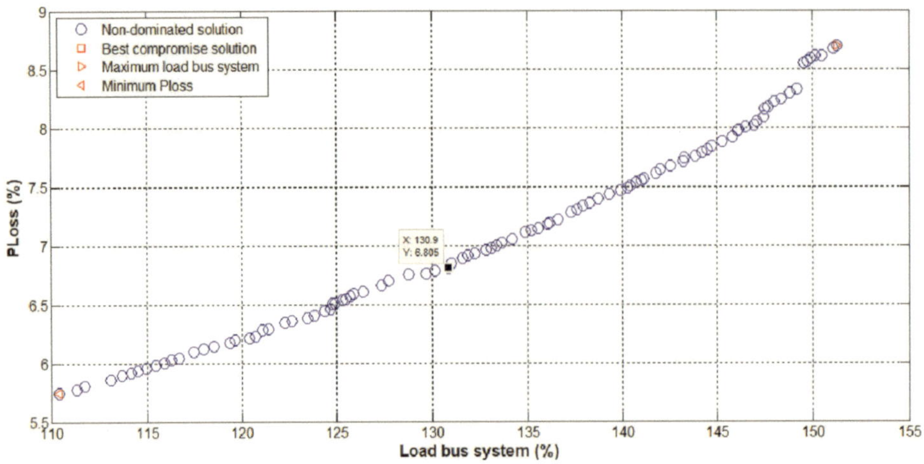

Figure 10.
Pareto front of Case-3.

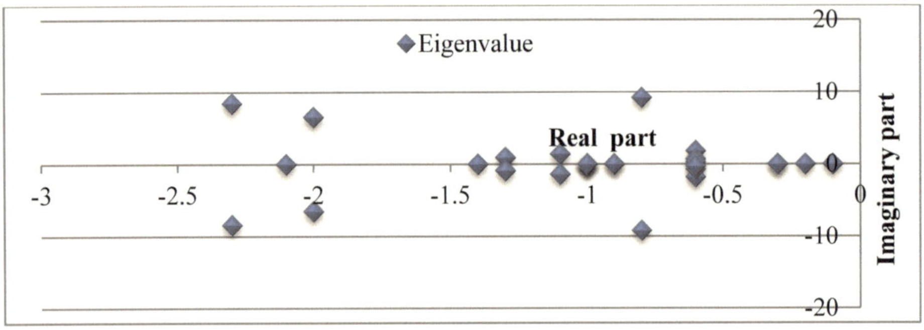

Figure 11.
The eigenvalue of Case-3.

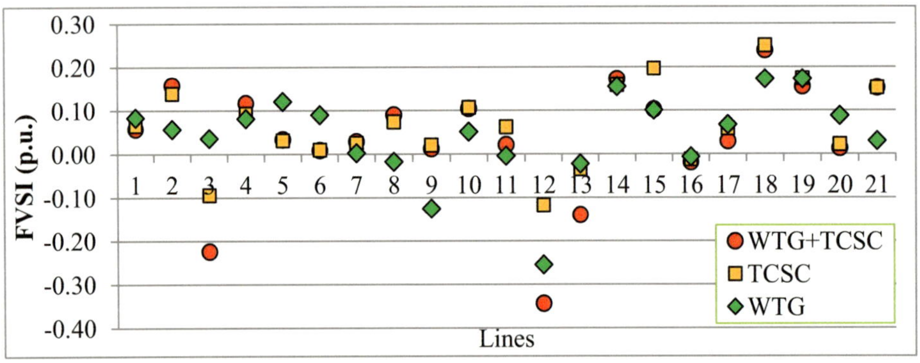

Figure 12.
FVSI of Case-3.

5.1.5 Case-4: SVC only

Figure 14 presents the Pareto front of the best solution obtained in Case-4 by placing a shunt FACTS controller, SVC, on busses 9 and 5 with a setting of 1.01 p.u.

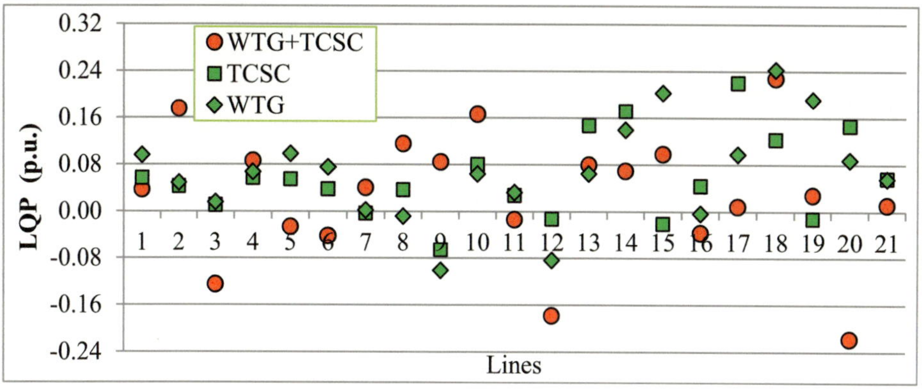

Figure 13.
LQP of Case-3.

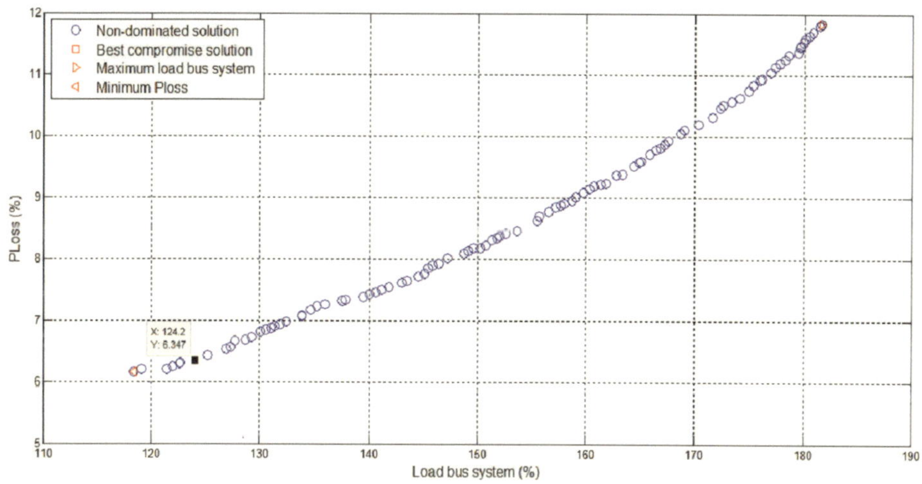

Figure 14.
Pareto front of Case-4.

and 1.09 p.u., respectively. Max. LBS and Min. P_{loss} reached 181.64% and 0.5562 p.u, respectively. Compared with the results of Case-1, the solution Min. P_{loss} and Max. The LBS obtained in this case was slightly increased. **Figure 15** depicts a small signal stability constraint satisfied by placing SVC on bus-13, with a setting of 1.05 p.u. and giving the best CS LBS and P_{loss} 124.16% and 0.2041 p.u, respectively.

5.1.6 Case-5: WTG and SVC

From the simulation results shown in **Figure 16** for Case-5, the best CS of LBS is 126.25% and P_{loss} 0.2429 p.u with the optimal placement of SVC on bus-7 by settings of 0.6144 pu and WTG on bus-3 with active power and reactive power of 39.15 MW and −23.20 MVAr, respectively. In addition, from the Figure, Max. LBS and Min. P_{loss} was 182.79% and 0.2107 pu, respectively, after placement of the shunt FACTS and WTG at their optimal locations. Compared to Cases 1 and 2, the best CS, in this case, is the highest. In all conditions, the eigenvalues of the stable system, as depicted in **Figure 17**.

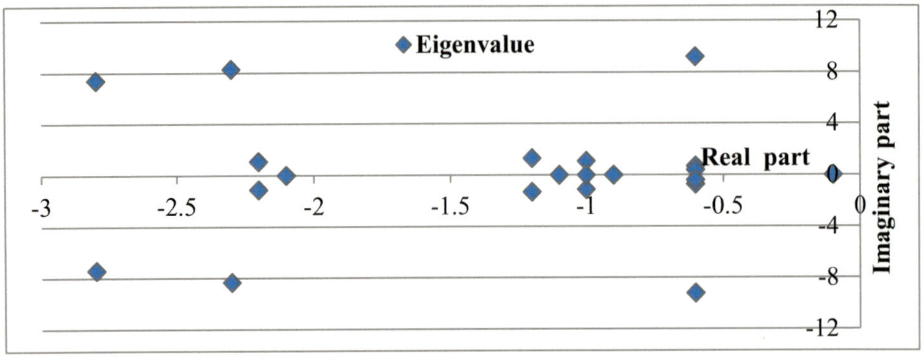

Figure 15.
Eigenvalue of Case-4.

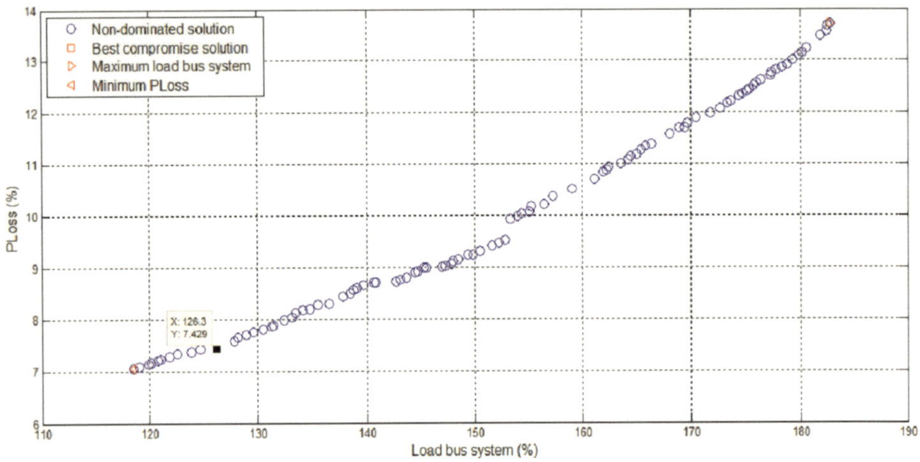

Figure 16.
Pareto front of Case-5.

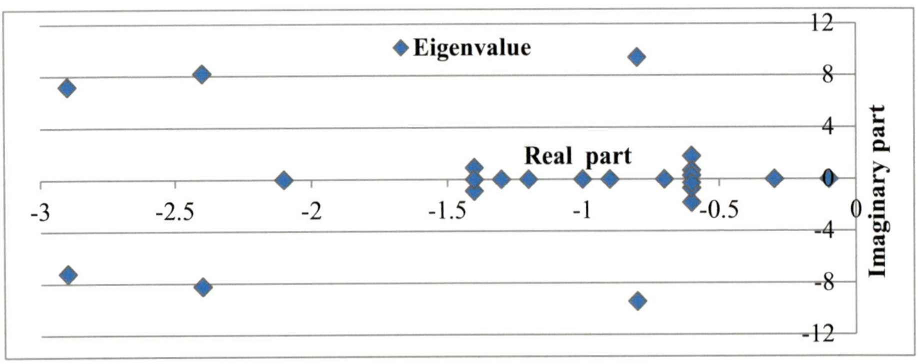

Figure 17.
Eigenvalue of Case-5.

In order to maintain the stability of the network voltage at numerous LBS levels and guarantee that no line is overloaded, then the two system security indices have been considered in this Case-5 as depicted in **Figures 18** and **19**. Both figures prove

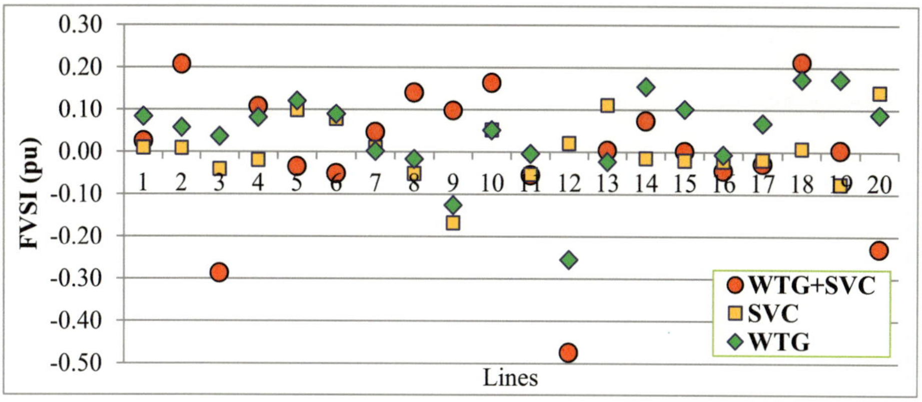

Figure 18.
FVSI of Case-5.

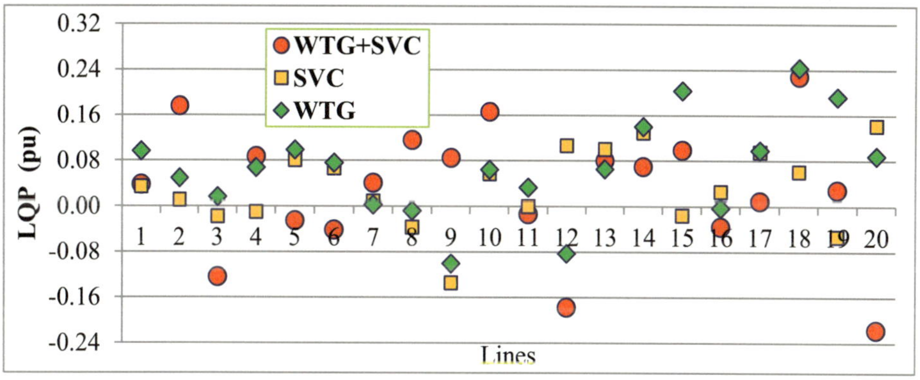

Figure 19.
LQP of Case-5.

that the FVSI and LPQ values for all cases, in this case, are less than 1.00, hence there is no bus collapse due to overload under any network conditions.

5.2 Practical test Indonesia Java-Bali 24-bus system

The developed method has also been successfully examined for its ability to modify the practical test system for 24 Java-Bali busses in Indonesia. Testing has been carried out to demonstrate more tangible results for all of the above case studies, but in this discussion, only the results are presented for Cases 3 and 5.

5.2.1 Case-3 for Java-Bali 24-bus system

The Pareto front result for Case-3 is depicted in **Figure 20**. This figure provides the integration of a WTG in the best locations on bus 12 with a size of 19.47 MW and −44.17 MVar and the optimal placing of a TCSC on lines 19–24 (GNDUL-NEWBLRJA) with the setting of −31.6923 p.u achieve the best CS results of

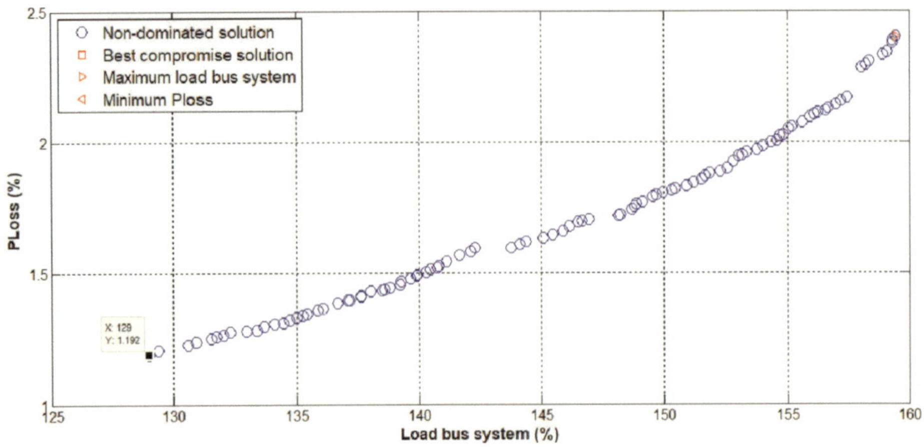

Figure 20.
Pareto front of Case-3 for Indonesia Java-Bali 24-bus system.

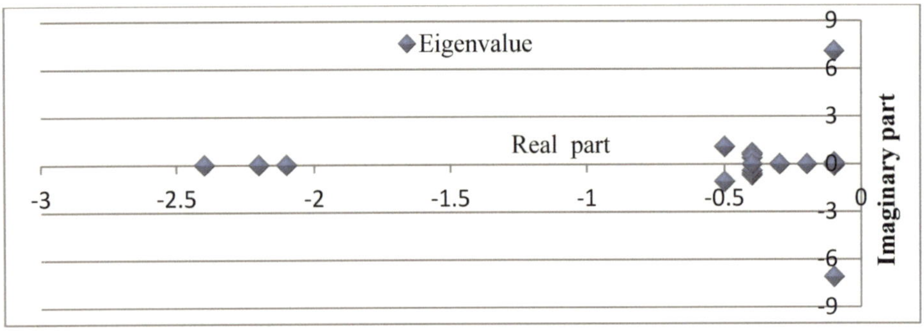

Figure 21.
The eigenvalue of Case-3 for Indonesia Java-Bali 24-bus system.

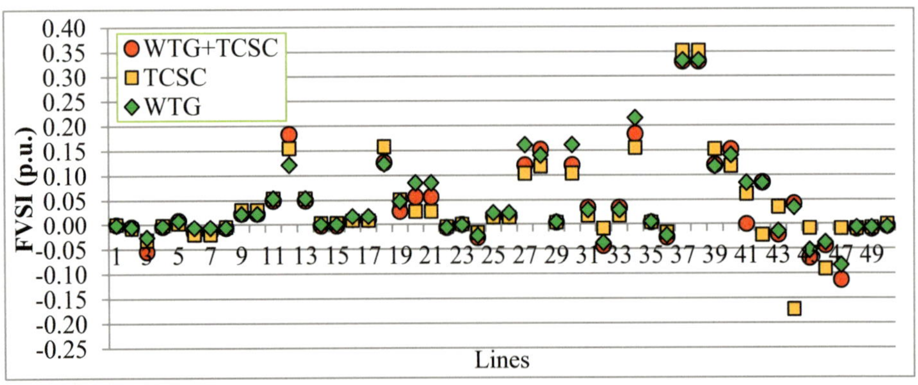

Figure 22.
FVSI of Case-3 for Indonesia Java-Bali 24-bus system.

optimal LBS and P_{loss} of 129% and 1.1917%, respectively. All the system stability constraints, in this case, small-signal stability, FVSI, and LQP, are satisfied, as described in **Figures 21–23**.

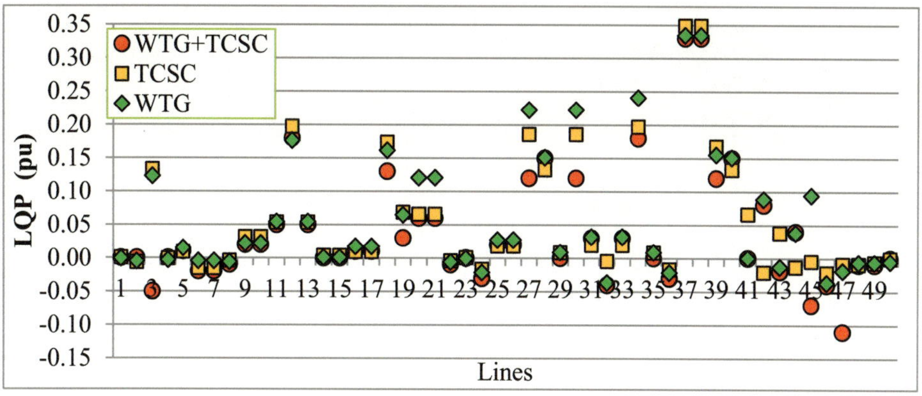

Figure 23.
LQP of Case-3 for Indonesia Java-Bali 24-bus system.

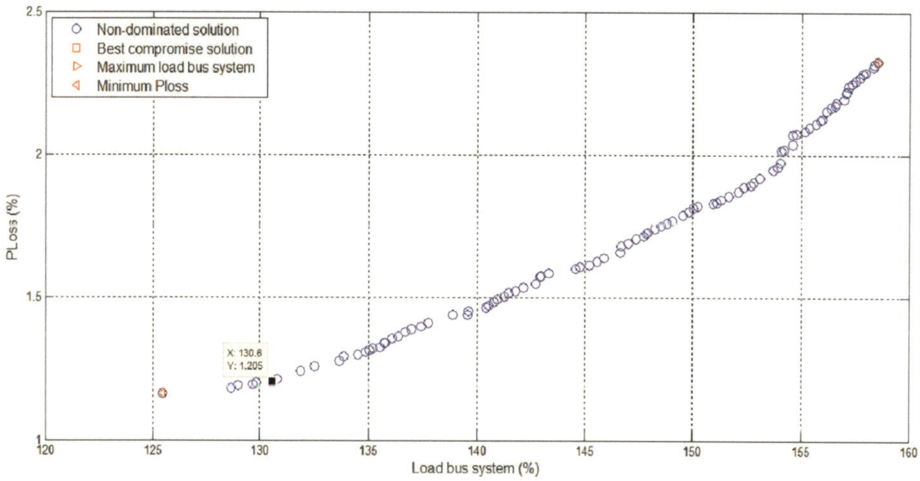

Figure 24.
Pareto front of Case-5 for Indonesia Java-Bali 24-bus system.

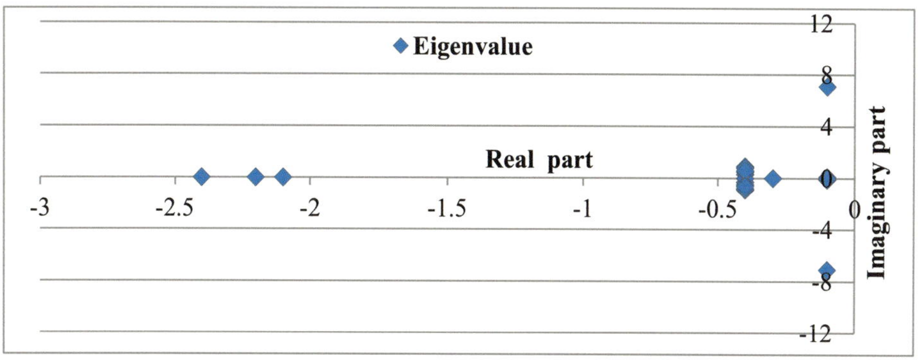

Figure 25.
The eigenvalue of Case-5 for Indonesia Java-Bali 24-bus system.

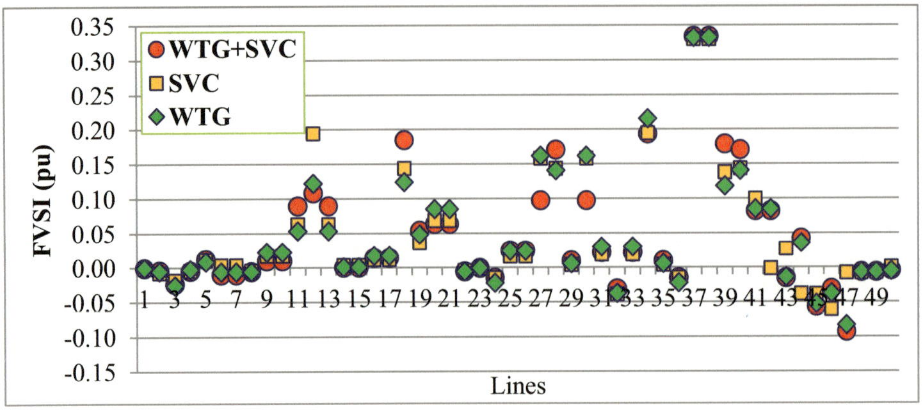

Figure 26.
FVSI of Case-3 for Indonesia Java-Bali 24-bus system.

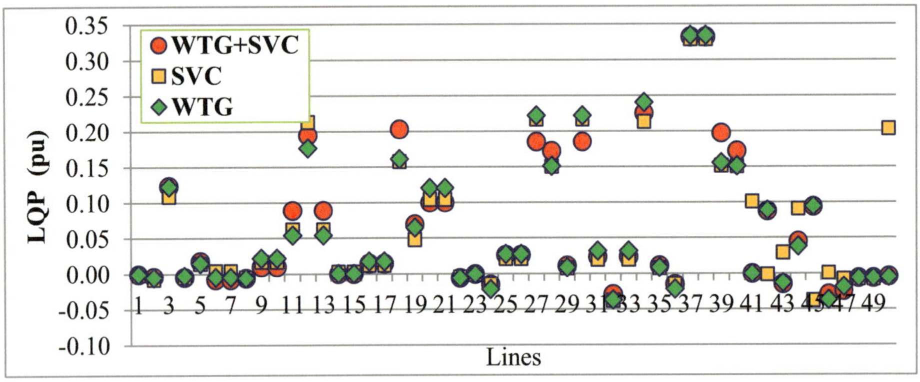

Figure 27.
LQP of Case-3 for Indonesia Java-Bali 24-bus system.

5.2.2 Case-5 for Java-Bali 24-bus system

The results of the Pareto front simulation for Case-5 on the Indonesian Java-Bali 24-bus system with the integration of a WTG and placement of an SVC are shown in **Figure 24**. From the figure, it can be observed that installing a WTG on bus-19 and an SVC on bus-13 and gives Max. SLB and Min. Ploss of 158.54% and 3.90 p.u, respectively. While the best CS LBS is 130.63% by placing an SVC in the same position but a WTG on bus-16. System stability and security based on system eigenvalues, FVSI, and LQP which prove the system is stable on the best CS, are presented in **Figures 25–27**.

6. Conclusions

Many system stability and security index constraints viz.: small-signal stability, FVSI, and LQP and status of system security constraints, that is, line thermal limit and bus voltage violation limits, have been successfully investigated. The indexes are satisfied with all the limitations in the DFIG type WTG integrated grid system by controlling two types of FACTS, TCSC and SVC. The problem formulated as a bi-optimization, maximizing the LBS system (Max. LBS), and minimizing P_{loss} (Min. P_{loss}) is both conflicting. To make it an optimal solution, the placing, and sizing of WTG and the setting of the series and shunt FACTS controllers are carried out using NSGA-II multi-objective optimization techniques. With the NSGA-II optimization techniques, the problem of bi-objective formulation has been solved to determine various cases of a combination of optimal locations, WTG (wind farm) size, and TCSC and SVC controller settings in the grid. The multi-objective approach has been successfully tested for effectiveness in the IEEE 14-bus standard test system and the Indonesia Java Bali 24-bus practical test system. In all cases, the proposed technique can keep all system stability constraints and safety margins within safe limits. Furthermore, a fuzzy-based mechanism quotes the best CS from the Pareto front.

Author details

I. Made Wartana* and Ni Putu Agustini
Department of Electrical Engineering, National Institute of Technology, Malang, Indonesia

*Address all correspondence to: m.wartana@lecturer.itn.ac.id

References

[1] Jeguirim M. Recent Advances in Renewable Energy Technologies. Elsevier; 2021

[2] Patel MR, Beik O. Wind and Solar Power Systems: Design, Analysis, and Operation. Boca Raton: CRC Press; 2021

[3] Ahmad J et al. Techno economic analysis of a wind-photovoltaic-biomass hybrid renewable energy system for rural electrification: A case study of Kallar Kahar. Energy. 2018;**148**: 208-234

[4] Răboacă MS et al. Concentrating solar power technologies. Energies. 2019; **12**(6):1048

[5] Nazir MS et al. Wind generation forecasting methods and proliferation of artificial neural network: A review of five years research trend. Sustainability. 2020;**12**(9):3778

[6] Paliwal P, Patidar N, Nema R. Planning of grid integrated distributed generators: A review of technology, objectives and techniques. Renewable and Sustainable Energy Reviews. 2014; **40**:557-570

[7] Shuaibu Hassan A, Sun Y, Wang Z. Optimization techniques applied for optimal planning and integration of renewable energy sources based on distributed generation: Recent trends. Cogentive Engineering. 2020;**7**(1): 1766394

[8] Rangu SK, Lolla PR, Dhenuvakonda KR, Singh AR. Recent trends in power management strategies for optimal operation of distributed energy resources in microgrids: A comprehensive review. International Journal of Energy Research. 2020; **44**(13):9889-9911

[9] Kim YJ, Kirtley JL, Norford LK. Reactive power ancillary service of synchronous DGs in coordination with voltage control devices. IEEE Transactions on Smart Grid. 2015;**8**(2): 515-527

[10] Selim A, Kamel S, Alghamdi AS, Jurado F. Optimal placement of DGs in distribution system using an improved Harris hawks optimizer based on single- and multi-objective approaches. IEEE Access. 2020;**8**:52815-52829

[11] Kumar NP, Rosalina KM. IPSO algorithm for maximization of system loadability, voltage stability and loss minimisation by optimal DG placement. System. 2015;**3**(11):73-77

[12] Kotsampopoulos P, Georgilakis P, Lagos DT, Kleftakis V, Hatziargyriou N. FACTS providing grid services: Applications and testing. Energies. 2019; **12**(13):2554:1-23

[13] Relić F, Marić P, Glavaš H, Petrović I. Influence of FACTS device implementation on performance of distribution network with integrated renewable energy sources. Energies. 2020;**13**(20):5516

[14] Gellings CW. The Smart Grid: Enabling Energy Efficiency and Demand Response. New York: River Publishers; 2020

[15] Liang X. Emerging power quality challenges due to integration of renewable energy sources. IEEE Transactions on Industry Applications. 2016;**53**(2):855-866

[16] Abdmouleh Z, Gastli A, Ben-Brahim L, Haouari M, Al-Emadi NA. Review of optimization techniques applied for the integration of distributed

generation from renewable energy sources. Renewable Energy. 2017;**113**: 266-280

[17] Shereen MA. Optimal allocation of DG units for radial distribution systems using genetic algorithm. International Journal of Engineering and Advanced Technology (IJEAT). 2012;**1**(6):175-179

[18] Farh HM, Al-Shaalan AM, Eltamaly AM, Al-Shamma AA. A novel crow search algorithm auto-drive PSO for optimal allocation and sizing of renewable distributed generation. IEEE Access. 2020;**8**:27807-27820

[19] Abdel-mawgoud H, Kamel S, Ebeed M, Aly MM. An efficient hybrid approach for optimal allocation of DG in radial distribution networks. In: 2018 International Conference on Innovative Trends in Computer Engineering (ITCE). IEEE; 2018. pp. 311-316

[20] Zad BB, Hasanvand H, Lobry J, Vallée F. Optimal reactive power control of DGs for voltage regulation of MV distribution systems using sensitivity analysis method and PSO algorithm. International Journal of Electrical Power & Energy Systems. 2015;**68**:52-60

[21] Kamarudin B, Hashim TJT, Musa A. Optimal Sizing and Location of Distributed Generation for Loss Minimization Using Firefly Algorithm. Indonesian Journal of Electrical Engineering and Computer Science. 2019;**14**(1):421-427

[22] Wartana I, Agustini NP, Singh JG. Optimal integration of the renewable energy to the grid by considering small signal stability constraint. International Journal of Electrical and Computer Engineering (IJECE). 2017;**7**(5)

[23] Nadjemi O, Nacer T, Hamidat A, Salhi H. Optimal hybrid PV/wind energy system sizing: Application of cuckoo search algorithm for Algerian dairy farms. Renewable and Sustainable Energy Reviews. 2017;**70**: 1352-1365

[24] Jamal T, Urmee T, Shafiullah G. Planning of off-grid power supply systems in remote areas using multi-criteria decision analysis. Energy. 2020; **201**:117580

[25] Islam MM, Muttaqi KM, Sutanto D. A novel saturated amorphous alloy core based fault current limiter for improving the low voltage ride through capability of doubly-fed induction generator based wind turbines. IEEE Transactions on Industry Applications. 2021;**57**(3): 2023-2034

[26] Mehta B, Bhatt P, Pandya V. Small signal stability analysis of power systems with DFIG based wind power penetration. International Journal of Electrical Power & Energy Systems. 2014;**58**:64-74

[27] Zhang C, Ke D, Sun Y, Chung C, Xu J, Shen F. Coordinated supplementary damping control of DFIG and PSS to suppress inter-area oscillations with optimally controlled plant dynamics. IEEE Transactions on Sustainable Energy. 2017;**9**(2):780-791

[28] Mohanty A, Viswavandya M, Mohanty S, Paramita P. Intelligent TCSC for enhancement of voltage stability and power oscillation damping of an off grid. In: Computational Intelligence in Data Mining. New Delhi: Springer; 2016. pp. 229-237

[29] Ziaee O, Choobineh FF. Optimal location-allocation of TCSC devices on a transmission network. IEEE Transactions on Power Systems. 2016; **32**(1):94-102

[30] Pati S, Dahiya R. Impact of wind integration on placement of TCSC. In: 2016 International Conference on Emerging Trends in Electrical Electronics & Sustainable Energy Systems (ICETEESES). IEEE; 2016. pp. 56-59

[31] Barrios-Martínez E, Ángeles-Camacho C. Technical comparison of FACTS controllers in parallel connection. Journal of Applied Research and Technology. 2017;**15**(1):36-44

[32] Agustini NP, Wartana IM, Lomi A. Improvement of static voltage stability of 16-bus bali system by optimal placement of SVC. International Journal of Smart Grid and Sustainable Energy Technologies. 2021;**4**(1):140-144

[33] Wartana IM, Singh JG, Ongsakul W, Sreedharan S. Optimal placement of FACTS controllers for maximising system loadability by PSO. International Journal of Power and Energy Conversion. 2013;**4**(1):9-33

[34] Wartana IM, Agustini NP. Application of Voltage and Lines Stability Index for Optimal Placement of Wind Energy with a System Load Increase Scenario

[35] Wartana IM, Singh JG, Ongsakul W, Buayai K, Sreedharan S. Optimal placement of UPFC for maximizing system loadability and minimize active power losses by NSGA-II. In: 2011 International Conference & Utility Exhibition on Power and Energy Systems: Issues and Prospects for Asia (ICUE). IEEE; 2011. pp. 1-8

[36] Milano F. An open source power system analysis toolbox. IEEE Transactions on Power Apparatus and Systems. 2005;**20**(3):1199-1206

[37] Pai M, Sen Gupta D, Padiyar K, Senroy N. Small Signal Analysis of Integrated Power Systems. New Delhi: Narosa Publishing House Pvt. Ltd.; 2016

[38] Agustini NP, Hayusman LM, Hidayat T, Wartana IM. Security and Stability Improvement of Power System Due to Interconnection of DG to the Grid. In: Proceedings of Second International Conference on Electrical Systems, Technology and Information 2015 (ICESTI 2015). Singapore: Springer; 2016. pp. 227-237

[39] Wartana IM. A multi-objective problems for optimal integration of the DG to the Grid using the NSGA-II. In: 2015 International Conference on Quality in Research (QiR). IEEE; 2015. pp. 106-110

[40] Deb K. A fast elitist non-dominated sorting genetic algorithm for multi-objective optimization: NSGA-2. IEEE Transactions on Evolutionary Computation. 2002;**6**(2):182-197

[41] Nicholas PE, Padmanaban K, Babu ML. Multi-objective optimization of laminated composite plate with diffused layer angles using non-dominated sorting genetic algorithm (NSGA-II). Advanced Composites Letters. 2014;**23**(4):403

[42] Agrawal S, Panigrahi BK, Tiwari MK. Multiobjective particle swarm algorithm with fuzzy clustering for electrical power dispatch. IEEE Transactions on Evolutionary Computation. 2008;**12**(5):529-541

[43] Zitzler E, Giannakoglou K, Tsahalis D, Periaux J, Papailiou K, Fogarty T. SPEA2: Improving the Strength Pareto Evolutionary Algorithm for Multiobjective Optimization. Vol. 952100. Barcelona: Proc of the EUROGEN Conf. Barcelona; 2002

Section 4

Renewable Energy Biofuels

Chapter 7

Oxyfuel Combustion in IC Engines

Jun Peng and Xiang Li

Abstract

This chapter introduces principles, mixing, ignition and combustion and controls processes of oxyfuel combustion which aims to achieve CCS (Carbon Capture and Storage) in IC (Internal Combustion) engines. By replacing air with pure oxygen and using hot and/or cooled EGR as dilutant gas for controlling the combustion process and flame speed, the mixing and combustion process will be explained. Fuel delivery, pre-mixing arrangement between pure oxygen and dilutant gas and their influences on combustion performances will be discussed. HCCI (Homogeneous Charge Compression Ignition), water injection, etc., technologies for enhancing the combustion efficiency will be demonstrated in detail. Finally, the emission characteristics and possible implementation of practical engine operation will be described.

Keywords: oxyfuel combustion, pure oxygen combustion, mixing, ignition, combustion process, combustion efficiency, EGR (Exhaust Gas Recirculation), CCS (Carbon Capture and Storage)

1. Introduction

As Net Zero Carbon Emissions strategy has been implemented in some countries and will be implemented gradually throughout the world, various zero carbon technologies are being developed, such as pure Battery Electric Vehicle (BEV) and Fuel Cell Vehicle (FCV) that have achieved certain mass production, although they have a high price and low driving range. For considering heavy-duty applications and with an alternative technology for a low-cost operation, hydrogen combustion technology and CCS (Carbon Capture and Storage) technology for hydrocarbon combustion are being developed widely [1].

It has been proposed that oxygen-enriched combustion and oxyfuel combustion are efficient ways to increase engine efficiency and reduce pollutant emissions. Oxyfuel combustion uses pure oxygen for combustion instead of air [2]. Due to the absence of nitrogen in the intake charge, NOx emissions will be eliminated. As a result, carbon dioxide and water vapour are the only products of combustion. Studies to date have been mainly focused on applying oxyfuel or oxygen-enriched combustion technologies to gas turbines and coal-fired power plants. The utilization of oxyfuel combustion and CCS for IC (Internal Combustion) engines has been gaining a lot of attention during the last few years [3]. Research on oxygen-enriched combustion shows that a slight increase in oxygen reduces smoke emissions as well as the amount of CO (Carbon Monoxide) and unburnt hydrocarbons but increases the amount of nitrogen oxides (NOx). Various technologies have been used to decrease NOx and particulates, such as Exhaust Gas Recirculation (EGR) and optimum injection strategies. Research conducted recently has drawn attention to oxyfuel and nitrogen-free combustion because of the benefits it brings to vehicles and is being used to make huge improvements to the efficiency of IC engines and to achieve zero NOx emissions. As is known, over-high peak pressures and peak pressure increases can easily appear in engine cylinders in oxygen-enriched conditions. An increase in combustion flame temperature is expected if oxygen is used instead of air. To minimize overheating problems due to overheating, it is crucial that the fuel injection flow rate should be accurately controlled in order to eliminate unexpected temperature rises in the premixed and diffusion combustion. Further, the diluent ratio and intake charge temperature have a meaningful impact on controlling in-cylinder temperature and combustion process.

Compared with other zero-carbon technologies being applied in various power and energy systems, oxyfuel combustion has the following advantages and drawbacks which should be paid attention.

Advantages:

- Oxyfuel combustion for being used on IC engines can make them to combining with CCS technology to achieve zero carbon emissions and other harmful emissions.

- Exhaust heat recovery can be easily integrated with oxyfuel combustion due to the requirement of steam condensation from the exhaust gas, and then, the total energy efficiency can be maintained.

- There is no nitrogen in the combustion mixture, and no NOx emission is produced from oxyfuel combustion.

Disadvantages:

- The requirement for oxygen from oxyfuel combustion will need some cost, although the cost can be reduced with low-cost oxygen supply from mass production water electrolysis for green hydrogen production for which oxygen is produced as a side product.

- As oxyfuel combustion needs diluent gas for slowing down the combustion speed and temperature and normally EGR (Exhaust Gas Recirculation) is selected for that. Then, CO_2 (Carbon Dioxide) in EGR will result in a reduction of thermal efficiency due to its high heat capacity compared to nitrogen.

- If oxyfuel combustion is employed for CCS, the CCS will also increase the whole system's cost.

2. Oxygen combustion principles

2.1 Principles

Pure oxygen getting combustion with conventional hydrocarbon fuels (liquid or gas) produces just carbon dioxide (CO_2) and water steam as combustion products. As water steam can be condensed to be liquid and using a liquid-gas separator to remove liquid water, only CO_2 gas is left and it can be stored, as demonstrated in **Figure 1**.

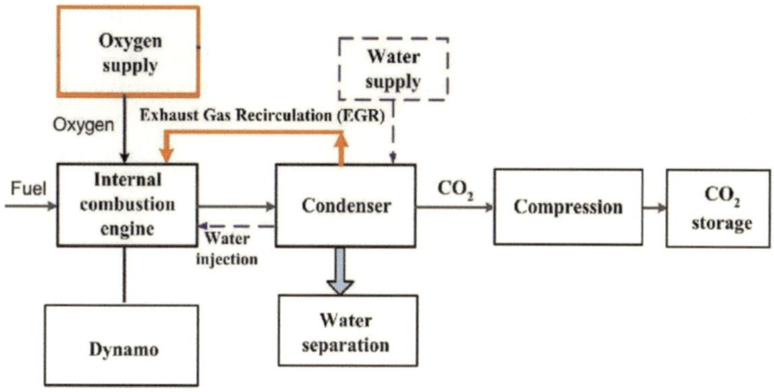

Figure 1.
Schematic of an application of oxyfuel combustion engine [1].

Because too fast reaction and too fast combustion flame propagation speed then result in too high combustion pressure increase rate and too high peak combustion temperature (although this will not produce high NOx emissions, while it lacks nitrogen in the mixture, it will need high-quality materials for the engine design), the combustion with the mixture of pure oxygen and hydrocarbon must be diluted. Basically, EGR (Exhaust Gas Recirculation) which consists of only CO_2 and water can be employed. Some research also used pure CO_2 or pure water (water injection).

Then, the stoichiometric reaction equation can be written as follows:

$$C_aH_b + \left(a + \frac{1}{4}b\right)O_2 + dilutant = aCO_2 + \frac{1}{2}bH_2O + dilutant \tag{1}$$

Replacing Air/Fuel Ratio (AFR) which is a very important control parameter used for conventional air-fuel combustion, Oxygen/Fuel Ratio (OFR) which should be employed for measuring the mixture for oxyfuel combustion can be defined as follows:

$$AFR = \frac{m_{air}}{m_{fuel}} \tag{2}$$

$$OFR = \frac{m_{oxygen}}{m_{fuel}} \tag{3}$$

For keeping a stoichiometric combustion (this is especially important for oxyfuel combustion while the stoichiometric combustion will save oxygen consumption) as demonstrated in Eq. (1), OFR can be derived from Eq. (1) as:

$$OFR_{Stoi} = \frac{m_{oxygen}}{m_{fuel}} = \frac{32\left(a + \frac{1}{4}b\right)}{(12a + b)} \tag{4}$$

If considering the normal air has 21% oxygen and 79% nitrogen by volumetric fraction or molar fraction, AFR for conventional combustion can also as expressed as function of OFR:

$$AFR = \frac{m_{air}}{m_{fuel}} = \frac{\left(m_{oxygen} + m_{nitrogen}\right)}{m_{fuel}} = \frac{\left(m_{oxygen} + \frac{79\% \times 28}{21\% \times 32} m_{oxygen}\right)}{m_{fuel}} = 4.29\frac{m_{oxygen}}{m_{fuel}} = 4.29OFR \tag{5}$$

As oxyfuel combustion has only water and CO_2 as the combustion products (the tiny fraction of other emissions can be ignored here), EGR (Exhaust Gas Recirculation) can be utilized as the dilutant for decelerating the reaction speed. The EGR rate can be expressed as:

$$EGR = \frac{m_{EGR}}{m_{EGR+Oxygen+fuel}} \quad (6)$$

GFR (Gas Fuel Ratio) for reflecting the ratio of in-cylinder all gas to fuel can be expressed as:

$$GFR = \frac{m_{oxygen+EGR}}{m_{fuel}} = \frac{(m_{oxygen} + m_{EGR})}{m_{fuel}} = OFR + \frac{m_{EGR}}{m_{fuel}} \quad (7)$$

2.2 Fuel, oxygen and dilutant supplies

For oxyfuel combustion engines, there is no difference for the fuel supply from conventional IC engines. Gasoline-oriented oxyfuel combustion engines can still have PFI (Port Fuel Injection) or GDI (Gasoline Direct Injection) whatever as used as the original design.

Although diesel engines can also operate oxyfuel combustion, it is not economic in terms of oxygen cost while it is not easy to keep stoichiometric oxyfuel combustion for diesel engines.

With regard to oxygen supply, pure oxygen can be supplied with oxygen tanks for light-duty applications, in particular those mobile light-duty applications. For heavy duty application, onboard oxygen production may be needed. Oxygen can be delivered to the oxyfuel engine through the intake manifold before or after mixed with diluent gas.

As mentioned previously, an easy option for dilutant is EGR. To delivery EGR gas into the cylinder, water, CO_2 and water + CO_2 should have appropriate mixing quality. A mixing chamber can be considered for this purpose.

For achieving ideal thermal efficiency, water is always a better candidate than CO_2 because water has a similar volumetric heat capacity as nitrogen but CO_2's volumetric heat capacity is much higher than nitrogen.

2.3 Combustion process and flame propagation

Replacing nitrogen as existing in air-fuel combustion with EGR (CO_2 or/and H_2O) as used in oxyfuel combustion, it results in a significant influence on combustion characteristics. As volumetric heat capacities of N_2, CO_2 and H_2O are 32.76 kJ/kmolK, 54.12 kJ/kmolK and 36 kJ/kmolK, respectively, the big increase of heat capacity from N_2 to CO_2 can lead oxyfuel combustion's flame speed obviously reduced under the same operating condition, compared with air-fuel combustion. Then, the combustion temperature and in-cylinder pressure will be reduced too much. As a result, oxyfuel combustion has lower thermal efficiency than air-fuel combustion if CO_2 is used as the only or main diluent gas. Therefore, for obtaining a better thermal efficiency, the dilutant of oxyfuel combustion should use as less as possible CO_2. As the increase of heat capacity from N_2 to H_2O is much smaller than that from N_2 to CO_2, EGR as main diluent choice for oxyfuel combustion can be treated for providing as high as possible H_2O fraction.

In a DNS (Direct Numerical Simulation) study presented by Zhong et al. [4] for oxyfuel combustion with CH_4 as fuel in a constant volume combustion chamber, it demonstrated the influence of CO_2 and H_2O on oxyfuel combustion characteristics. As shown in **Table 1**, five modelling cases were tested for Case 1—with N_2 (Nitrogen) as dilutant gas (or conventional air-fuel combustion), Case 2—with H_2O (steam) as dilutant gas, Case 3—CO_2 as dilutant gas, Case 4—with H_2O as dilutant but its chemical effects are isolated (FH_2O), Case 5—with CO_2 as dilutant gas also its chemical effects are isolated (FCO_2). The method for isolating chemical effects of H_2O in Case 4 or CO_2 in Case 5 is to stop those reactions between pre-added H_2O or CO_2 and other species. Those H_2O and CO_2 produced during the combustion process are still allowed to have reactions with other species.

In **Figure 2**, it can be seen that CO_2 additive (Case 3) can significantly decelerate the combustion speed and then slow down the increase of combustion temperature very obviously, compared with air-fuel combustion (Case 1). H_2O additive (Case 2) does not reduce the combustion speed, compared with Case 1, and even accelerates that a little. This suggests that using pure water steam as dilutant gas in oxyfuel combustion would keep the flame speed, combustion temperature and thermal efficiency very similar to air-fuel combustion. But CO_2 will provide a significant disadvantage for those.

In the same study, by isolating chemical effects of H_2O (Case 4) and CO_2 (Case 5), it can be found that CO_2 additive even can remarkably accelerate the combustion speed. That suggests CO_2's chemical effects play much bigger role than its thermal effects in decelerating the combustion speed. Water steam's chemical effects have a

Case	CH_4 (%)	O_2 (%)	N_2 (%)	CO_2 (%)	H_2O (%)	FCO_2 (%)	FH_2O (%)	S_L (m/s)	u'/S_L	T_b (K)
1	5.5	22	72.5	0.0	0.0	0.0	0.0	1.02	2.36	2537
2	9.0	36.1	0.0	0.0	54.9	0.0	0.0	1.11	2.17	2508
3	6.4	25.5	0.0	68.1	0.0	0.0	0.0	0.69	3.49	2535
4	9.0	36.1	0.0	0.0	0.0	0.0	54.9	1.08	2.22	2552
5	6.4	25.5	0.0	0.0	0.0	68.1	0.0	1.18	2.04	2649

Table 1.
DNS simulation parameters and results [4].

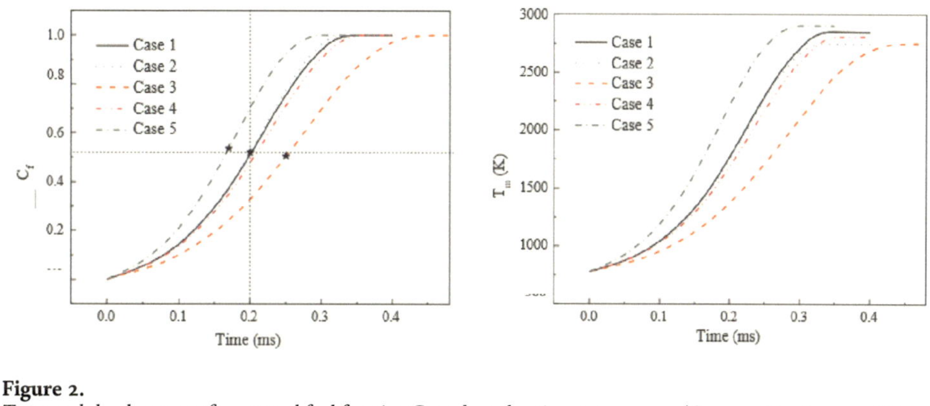

Figure 2.
Temporal development of consumed fuel fraction C_f and combustion temperature (Case 1—N_2, Case 2—H_2O, Case 3—CO_2, Case 4—FH_2O, Case 5—FCO_2) [4].

reverse function, and the magnitude is very small. In **Table 1**, laminar flame speeds (S_L) and other results of five cases are also listed.

In an experimental study of partially premixed CH_4/oxyfuel flames in a swirl stabilised burner [5], it demonstrated that the shape and stabilisation of oxyfuel combustion flame are closely related to the oxygen concentration, and it was suggested that the oxygen fraction in the mixture of oxygen and dilutant gases should be at least similar as in air (volumetric 21%) for maintaining the stabilisation of oxyfuel combustion flames.

3. Oxygen combustion engines and performances

3.1 Gasoline-oriented oxyfuel combustion

So far, most reported oxyfuel IC engines are gasoline-oriented engines. Because gasoline engines can be easily operated under stoichiometric conditions, they will not waste oxygen, which results in a certain cost for oxyfuel combustion. As shown in **Figure 3** below, a gasoline engine modified for oxyfuel combustion which was used by Wu et al. [6] employs almost the same configuration as the core engine combustion system. The fuel delivery is conventional PFI (Port Fuel Injection), and the gas path has a pure oxygen supply from an oxygen tank. CO_2 also from a tank is supplied to the intake manifold for representing EGR. For increasing water content in dilutant gas, a water injection system is fitted for directly inject water into the cylinder.

By experimental investigation, in-cylinder pressure traces are recorded for demonstrating the effects of water injection on the increase of in-cylinder pressure [7], as shown in **Figure 4**. The in-cylinder traces show that water injection can remarkably increase in-cylinder pressure during expansion stroke. If the water injection time is adjusted for the optimal crank angle, a very ideal thermal cycle efficiency can be achieved. The same researchers have carried out comprehensive for applying water injection to improve oxyfuel combustion's thermal efficiency and optimal operating and control parameters have been presented.

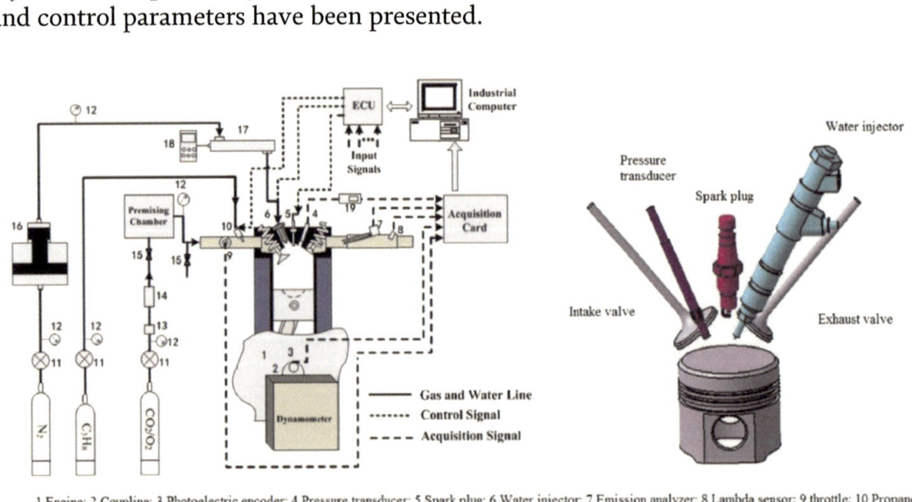

1 Engine; 2 Coupling; 3 Photoelectric encoder; 4 Pressure transducer; 5 Spark plug; 6 Water injector; 7 Emission analyzer; 8 Lambda sensor; 9 throttle; 10 Propane nozzle; 11 Pressure reduction valve; 12 Pressure gage; 13 Throttle valve; 14 Flowmeter; 15 Ball valve; 16 Pressure accumulator; 17 Water rail; 18 thermocouple; 19 charge amplifier

Figure 3.
Gasoline oxyfuel combustion engine with Direct Water Injection (DWI) [6].

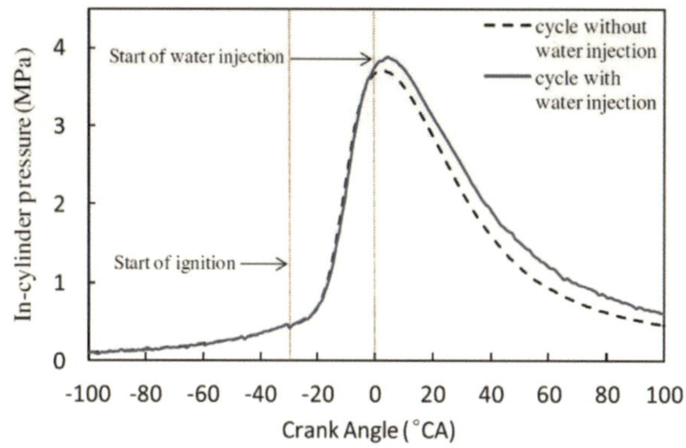

Figure 4.
In-cylinder pressure traces demonstrating the influence of DWI [7].

In **Figure 5**, effects of water injection temperature on the thermal efficiency of oxyfuel combustion are presented. The results suggest that the increase of water injection mass will linearly increase the thermal efficiency. Meanwhile, higher water temperature will always benefit the thermal efficiency too. With a heat exchanger for allowing injected water heated from exhaust heat, the water temperature up to 200°C can be achieved [8].

Effects of intake pressure with increased water injection mass are demonstrated in **Figure 6**. The increase of intake pressure can be understood for meeting the requirement of engine power increase, while fuel amount can increase with the increase of intake pressure. As shown in **Figure 6**, increased water to total intake mass ratio can result in MEP (Mean Effective Pressure) increase with various intake pressure. The higher the intake pressure is, the bigger increase of MEP by water to total intake mass ratio provides.

3.2 Diesel-oriented oxyfuel combustion

Diesel-oriented oxyfuel combustion engines have been paid attention to because they can be operated with much heavier duty than gasoline engines. Most applications with IC engines integrated with oxyfuel combustion and CCS (Carbon Capture and Storage) are those heavy engines, such as those used in maritime sector and stationary power generation sector.

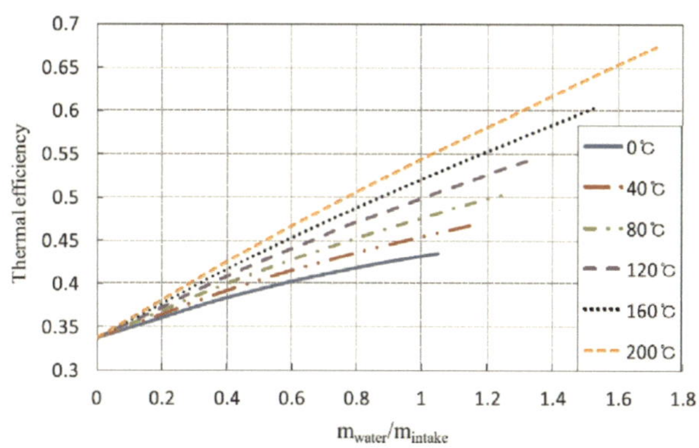

Figure 5.
Effects of water injection temperature on thermal efficiency [8].

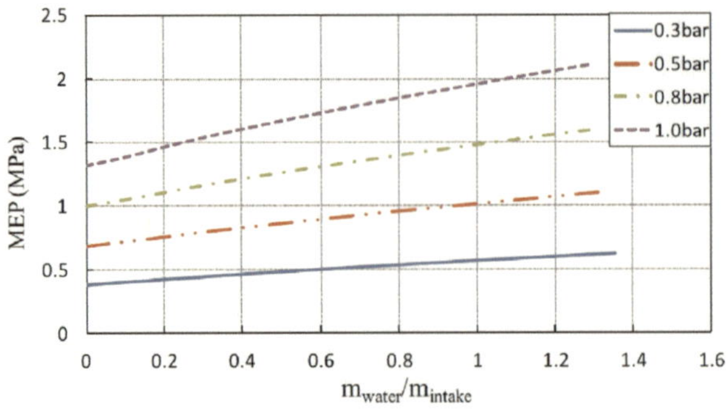

Figure 6.
Effects of intake pressure on MEP [8].

Although it is difficult to operate diesel engines under stoichiometric conditions, optimal combustion system configuration and control can make the OFR (Oxygen/Fuel Ratio) very close to the stoichiometric condition, and then, wasted oxygen amount can be limited.

Because diesel combustion needs high AFR (OFR for oxyfuel diesel combustion), the combustion process is very sensitive to oxygen concentration in the mixture. As shown in **Figure 7**, reduced oxygen concentration (when the total mixture amount in the cylinder is constant) can obviously reduce in-cylinder pressure (then definitely combustion temperature). As a result, IMEP and thermal efficiency will reduce with the same fuel injection amount [9, 10].

For reducing the cost of oxygen supply, it is always beneficial to have as low as possible OFR (Oxygen/Fuel Ratio). But reduced oxygen concentration can result in the mixture contaminated with inadequate oxygen for achieving complete combustion for diesel oxyfuel combustion. Reflected on ISFC (Indicated Specific Fuel Consumption), stoichiometric OFR with the lowest ISFC can be found with adjustable oxygen to CO_2 ratio, as shown in **Figure 8**.

In **Figure 8**, Lambda$_{O2}$ (relative OFR) is defined as:

$$Lambda_{O2} = \frac{OFR_{actual}}{OFR_{stoi}} \tag{8}$$

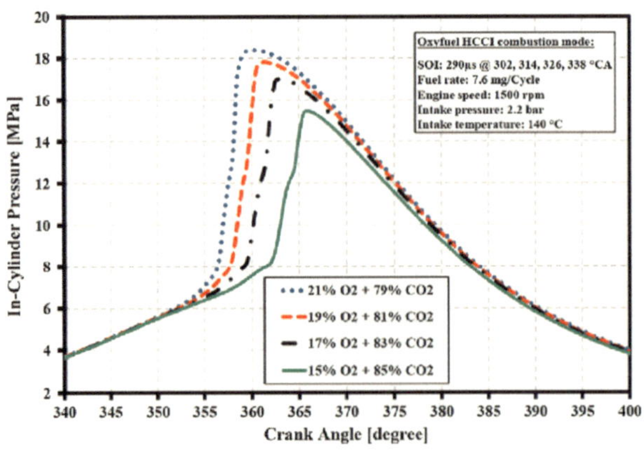

Figure 7.
Effects of oxygen concentration on diesel oxyfuel combustion process [9].

In **Figure 8**, at Lambda$_{O2}$ = 1.0, the combustion with stoichiometric mixture is the most economic point with regard to the cost of oxygen supply. Achieving a lower ISFC (Indicated Specific Fuel Consumption) but maintaining Lambda$_{O2}$ at 1.0 will be desired. Then, lower oxygen concentration with higher total in-cylinder mass due to increased dilutant gas CO_2 (green line in **Figure 8**) is a better choice than other conditions.

Intake temperature also needs carefully adjusted, while it also influences diesel oxyfuel combustion performances remarkably. As shown in **Figure 9**, higher intake temperature can reduce IMEP (Indicated Mean Effective Pressure) because higher intake temperature results in less intake gas amount, and then less fuel amount and lower IMEP.

In **Figure 10**, it shows that the reduced IMEP by increased intake temperature may be more due to the advanced ignition timing. As the intake temperature increases from 140°C (blue line in **Figure 10**) to 220°C (yellow line), the ignition timing can advance from about 355 CA (Crank Angle) to about 345 CA. Earlier ignition timing makes the heat release taking place too early before TDC will no doubt increase compression negative work, then IMEP reduces.

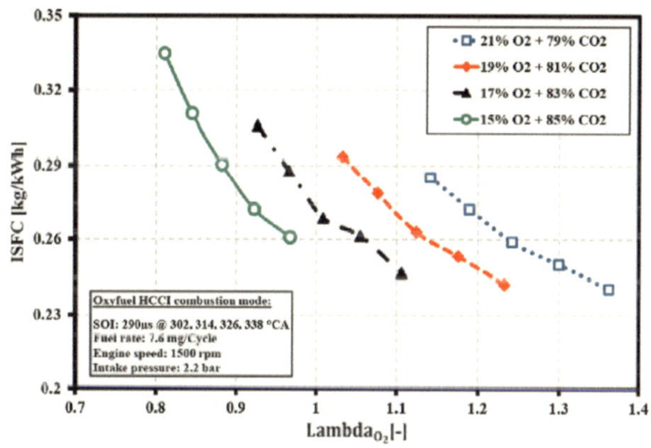

Figure 8.
Effects of relative OFR (Lambda$_{O2}$) on ISFC [9].

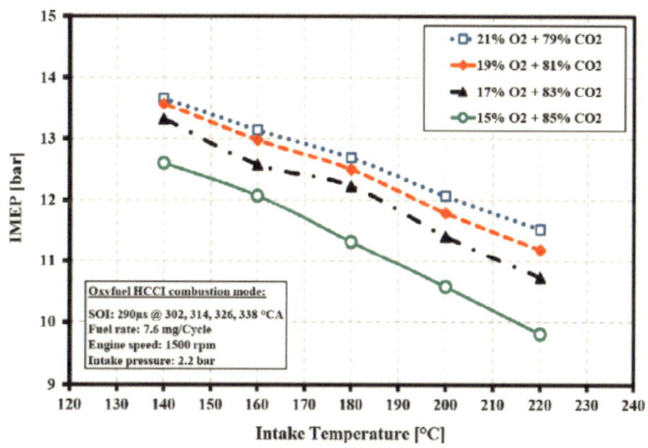

Figure 9.
Effects of intake temperature on IMEP under different oxygen concentrations [10].

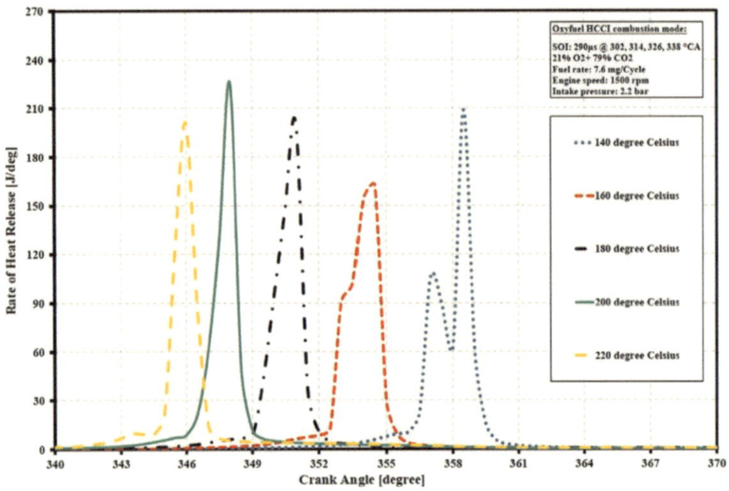

Figure 10.
Effects of intake temperature on ignition timing [10].

From the viewpoint for a better mixing, late ignition timing with the same injection timing will lead long ignition delay, which can result in more homogeneous mixture with long fuel evaporation and mixing time. Then, HCCI (Homogeneous Charge Compression Ignition) combustion under oxyfuel combustion mode can be achieved. Several reported studies of diesel oxyfuel combustion [9–11] have demonstrated their explorations into HCCI oxyfuel combustion in diesel engines.

Because higher intake temperature can result in lower IMEP under the same fuel amount, ISFC with increased intake temperature has obvious increase, as shown in **Figure 11**. Those results suggest that operating parameters for oxyfuel combustion play very important role for influencing oxyfuel combustion engine performance, similar as conventional air-fuel combustion. Careful calibration should always be necessary for achieving desired engine performances.

By experimental investigation, diesel oxyfuel combustion has been examined by several report research. Reported by Kang et al. [11] for studying diesel oxyfuel

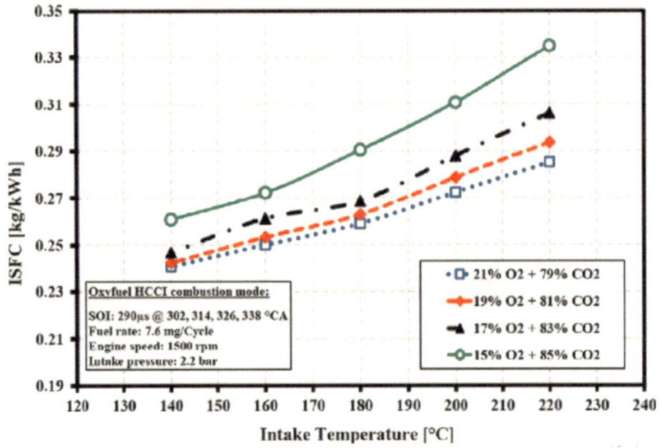

Figure 11.
Effects of intake temperature on ISFC [10].

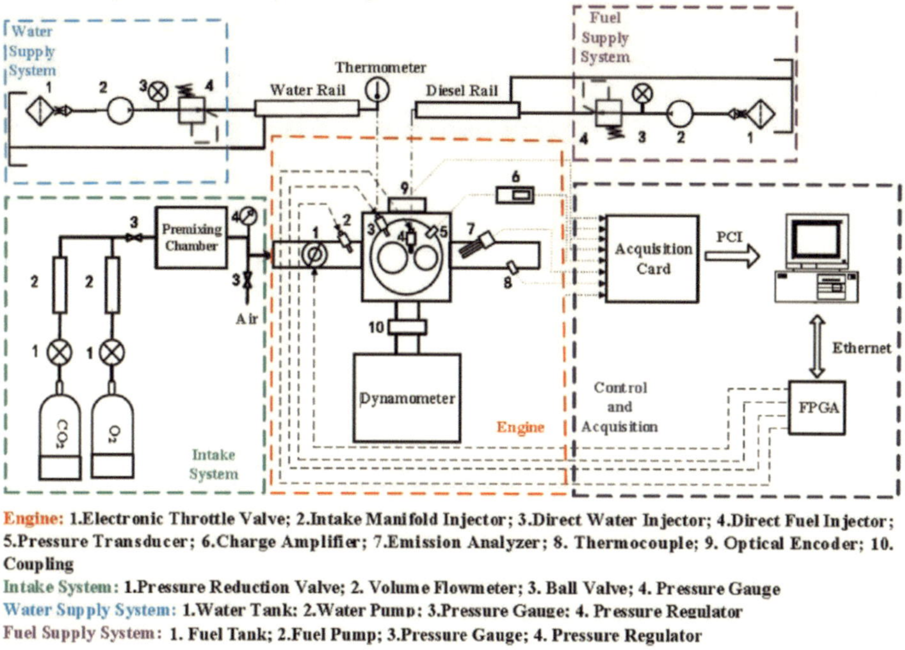

Figure 12.
A diesel test engine modified to operate oxyfuel combustion [11].

combustion in a test diesel engine, conventional common rail high-pressure injection system is employed, as shown in **Figure 12** below. By arranging CO_2 and O_2 supply from gas tanks and with direct water injection for achieving higher thermal efficiency, the effects of water injection temperature on in-cylinder pressure can be found in **Figure 13** below. From the result, it can be seen that water injection can result in a small reduction of compression negative work, a small reduction of expansion work, and an obvious increase of ignition delay and retarded ignition timing. By increasing

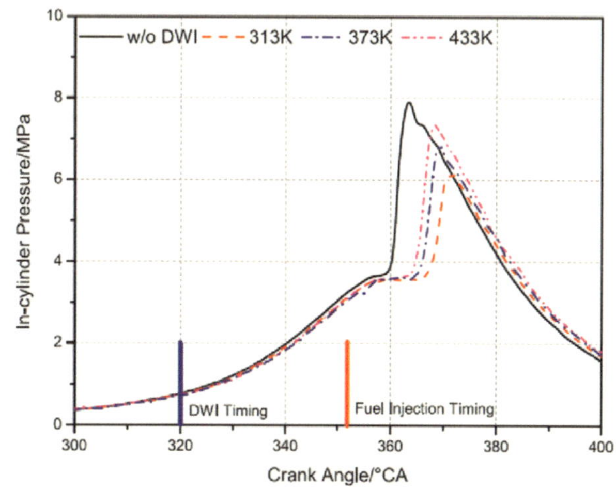

Figure 13.
In-cylinder pressure traces under different water injection temperature [11].

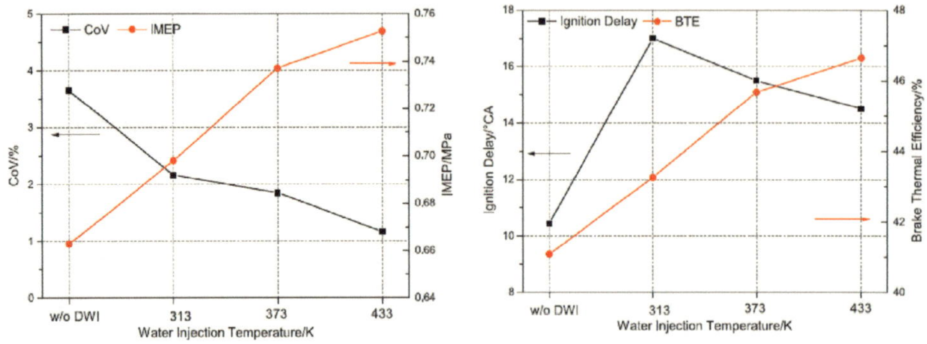

Figure 14.
Effects of water injection temperature on IMEP, COV of IMEP, ignition delay and thermal efficiency [11].

the water temperature, it shows that expansion work can obviously increase, while the compression negative work has no obvious variation.

With the combustion analysis based on the measured in-cylinder pressure traces, the experiment work reported that the increased water temperature can benefit oxyfuel combustion performances significantly.

As shown in **Figure 14**, based on the reported operation condition, IMEP (Indicated Mean Effective Pressure) gets some increase and COV (Coefficient of Variation of IMEP) gets reduction. Finally, the increase of water injection temperature makes thermal efficiency increased.

3.3 Emissions

As discussed previously, oxyfuel will not produce NOx emissions due to the absence of nitrogen in the mixture. Although the combustion can result in uHC (unburnt Hydrocarbon) and CO emissions, those can be captured and storage with CO_2 together. Therefore, generally there is not concern of emissions for oxyfuel combustion.

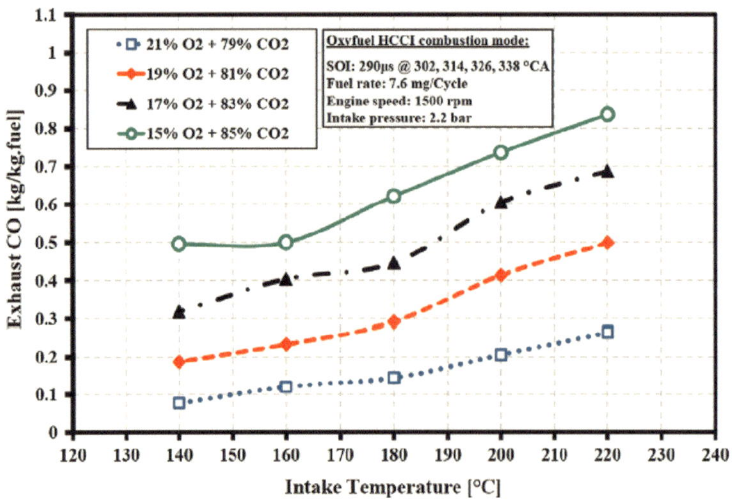

Figure 15.
Effects of intake temperature on CO emissions [10].

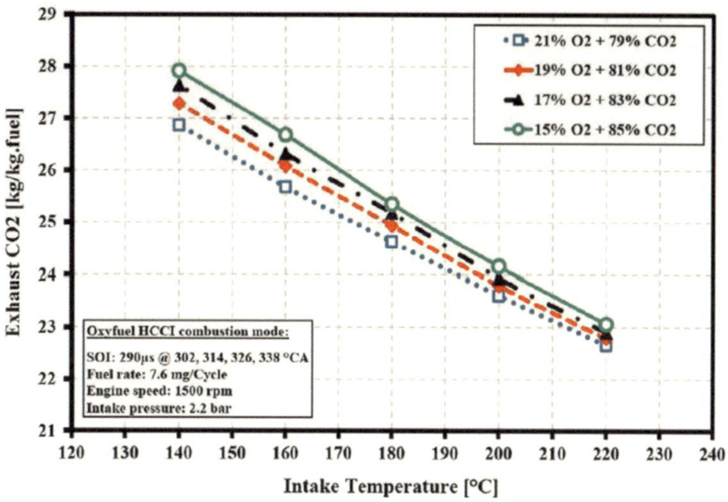

Figure 16.
Effects of intake temperature on CO_2 emissions [10].

Therefore, the combustion efficiency should be always given priority over emissions for oxyfuel combustion process optimization. Reported level of CO emissions from oxyfuel diesel combustion by Mobasheri et al. [10] can be up to 0.8 kg/kgfuel under 15% of oxygen concentration and 220°C of intake temperature from less than 0.1 kg/kgfuel under 21% of oxygen concentration and 140°C intake temperature, as shown in **Figure 15** below.

Compared with CO_2 level (shown in **Figure 16**) in the same exhaust gas which are about 23 kg/kgfuel and 27 kg/kgfuel under those two operating conditions, it can be estimated that incomplete combustion (just considering CO from the incomplete combustion but not including uHC) can be between 0.3% and 3.5%. This means that some amount energy is not fully released during combustion stage. As CO is mainly produced from inadequate oxygen in the mixture, increasing OFR will be necessary for diesel oxyfuel combustion, if the price of oxygen is not a big problem.

4. Oxygen productions and supplies

Supplying oxygen to oxyfuel combustion can be carried out in several ways. The first one is to use high-pressure oxygen tanks. This is more suitable for mobile and light-duty applications. The second alternative is using an oxygen generator working on the swing adsorption technique using pump compressors and air driers for producing oxygen with a purity of approximately 95%. This will be suitable for mobile but heavy-duty applications. For those heavy-duty but stationary applications, oxygen produced from water electrolysis with mass production can be considered. As green or blue hydrogen production are being developed for meeting the requirement of Net Zero Carbon Emissions, hopefully oxygen as a side product will become abundant in the near future.

Considering useful energy output of 1 kg gasoline or diesel fuel from oxyfuel combustion is about 44 MJ × 40% = 17.6 MJ. The gasoline or diesel fuel of 1 kg consists of about 0.87 kg of carbon (or 0.87/12 = 0.0725 kmol) and 0.13 kg of H_2 (or 0.13/2 =

0.065 kmol). To fully burning those carbon and hydrogen, the required oxygen amount will be:

$$0.0725C + 0.065H_2 + 0.105O_2 = 0.0725CO_2 + 0.065H_2O \quad (9)$$

$$0.105\ kmol \times 32 = 3.36\ kg \quad (10)$$

Therefore, 1 kg of gasoline or diesel fuel for complete combustion will require about 0.105 kmol or 3.36 kg of oxygen.

4.1 Oxyfuel tank

If the oxygen supply is supplied by commercial oxygen cylinder for lab application, the current price from BOC Company in the UK is about £2/kg. Burning 1 kg gasoline or diesel fuel and 3.36 kg oxygen will require £6.72. Using the current diesel price £2/litre or £2.44/kg in the UK market, the cost for oxygen is nearly three times of diesel fuel price. This is too expensive if using commercial lab oxygen supplied from BOC.

By checking possible lowest price for oxygen, it is found that in China, 99.5% of oxygen for industrial applications can be down to RMB 2.0/kg, equivalent to £0.22/kg. If similar technology can be provided in the United Kingdom and further improvements can be made for mass production, the cost for oxygen supply by cylinder will be acceptable.

Using commercial oxygen supply, safety regulations for handling and use of high-pressure gas cylinders must be strictly adhered to for use and installation on oxyfuel combustion engines. In this case, the EU pressure equipment directive would be applicable to the onboard use of oxygen gas cylinders. More information about this regulation can be found online [12].

If high-pressure oxygen gas cylinders are used on oxyfuel combustion engines, other components such as pressure regulators, electronic flow control valves and piping to engine are required. A flow control circuit that is controlled by the engine speed requirements must be designed and installed to provide the right amount of oxygen at the engine intake as required.

4.2 Onboard oxyfuel production

Cryogenic Compression Cycle: This technology can separate oxygen from air using a cryogenic compression cycle [13]. In **Figure 17**, it illustrates some major unit

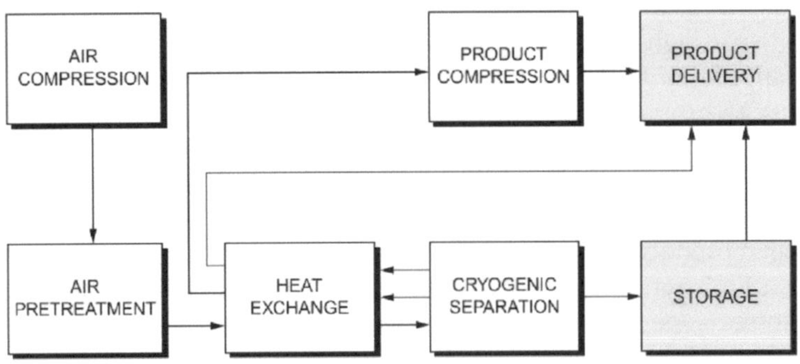

Figure 17.
Cryogenic compression cycle system [13].

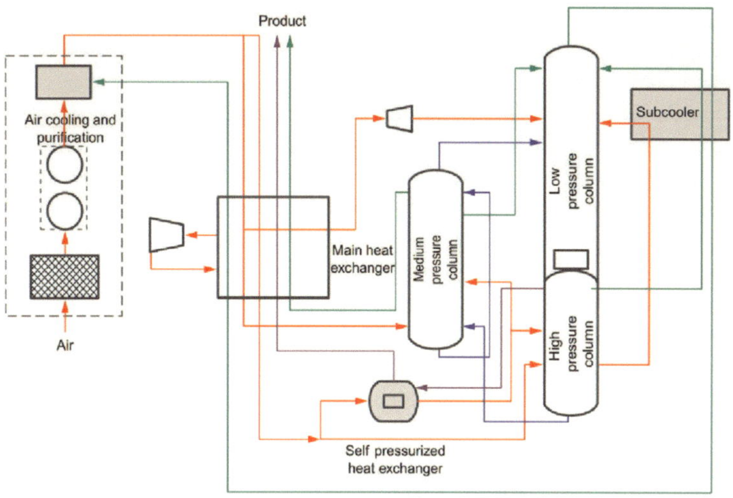

Figure 18.
Schematic of a cryogenic distillation technology [14].

operations required to cryogenically separate air into useful products. The method will at first lower the temperature of air until it liquefied. Then, liquid air is distilled, and the component gases are separated from each other. Each component that boiled off was separated and captured individually. This cryogenic distillation process was the foundation for the industrial production of high-grade oxygen, nitrogen and argon.

Cryogenic Distillation: As demonstrated in **Figure 18**, the cryogenic distillation technique [14] can be used when either very high-quality oxygen (>99.5%), high volumes of oxygen (≥102 tons of oxygen/day) or high-pressure oxygen are required.

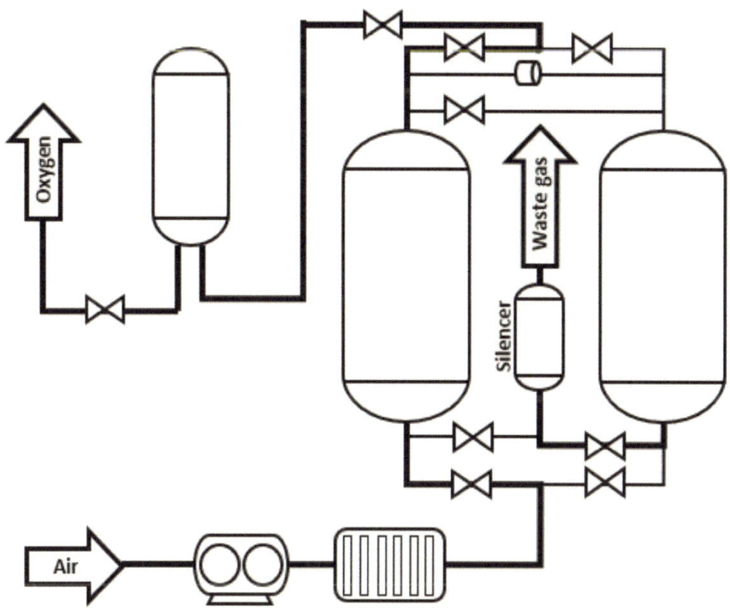

Figure 19.
Schematic of pressure swing adsorption [15].

Cryogenic air separators need to take more than an hour to start up. Additionally, since cryogenics can produce such a high purity of oxygen, the waste nitrogen stream is of usable quality. This can add significant financial benefits to a process integrated with a cryogenic. However, the cryogenic distillation process and equipment are very expensive and suited for large industrial oxygen production plants.

Pressure Swing Adsorption (PSA): Pressure Swing Adsorption (PSA) technology, as demonstrated in **Figure 19**, is the most promising from an energy and cost point of view. The technology involves an adsorption/desorption of gases onto solid surfaces while under pressure [15]. In fact, the pressurising of the system means that the solid material can adsorb more gases than under normal atmospheric conditions. The gases are usually pressurised to about 8 bar and adsorbed on the solid materials. Different solid substrates attract different gases; hence, the oxygen can be separated from air using this technique. This pressure adsorption technique is also being investigated for adsorption and storage of CO_2 from the exhaust in thermal power stations as post-combustion CCS. To produce 1 m^3 of oxygen with the pressure, swing adsorption technology needs about 1 kWh of energy.

5. Implementation of oxyfuel combustion

5.1 Case Study I—Gasoline oxyfuel combustion with water injection

As shown in **Figure 20**, a practical system for implementing gasoline oxyfuel combustion with water injection was developed by Wu et al. [6]. The system has a PFI fuel injection system. The gas path has a pure oxygen supply from a liquid tank. Before the oxygen is fed into the intake manifold to mix with EGR, a heat exchange utilities exhaust heat warming the oxygen flow. After the EGR manifold, exhaust gas is firstly cooled with a heat exchanger for condensing steam and then separating liquid water. The liquid water then is reheated through the coolant channel of the engine

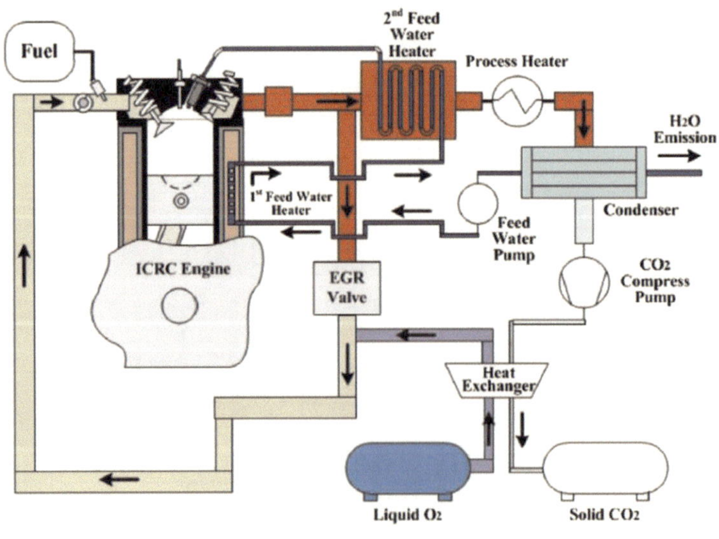

Figure 20.
Gasoline oxyfuel combustion engine [6].

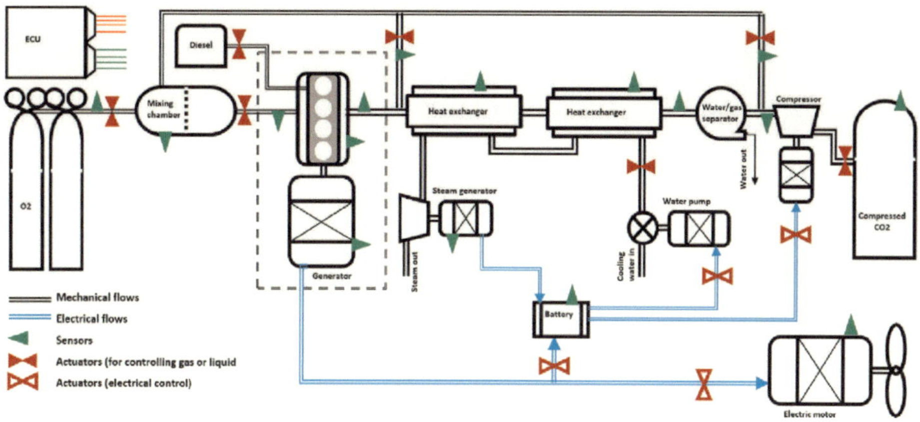

Figure 21.
Gasoline oxyfuel combustion engine [16].

block and the heat exchanger. Finally, the heated water is injected into the cylinder directly.

The advantages of this configuration can have as more as possible water and as less as possible EGR (consisting of CO_2). The water temperature can also be adjusted by controller by controlling the water flow or by a bypass valve outside the heat exchanger. Then, the oxyfuel combustion performance can be maintained at satisfactory level.

5.2 Case Study II—Oxyfuel combustion engine for inland boats

This case is about a practical application of diesel oxyfuel combustion engine on an inland boat where there is possible space for installing the exhaust gas after treatment system [16, 17]. As shown in **Figure 21**, the oxygen supply has no difference from that as used in Case Study 1. Diluent gas mainly comes from raw EGR and/or treated EGR, depending on the requirement for controlling the combustion performance. Two stages of heat exchangers are employed for ensuring an adequate condensation of steam in exhaust gas. An exhaust heat recovery cycle based on Rankine cycle is integrated for increased the whole system's energy efficiency. Liquid water in condensed exhaust gas is separated with a water-gas separator; then, CO_2 is compressed and stored in an onboard tank.

6. Summary

Although oxyfuel combustion has been studied on gasoline and diesel engines by a number of researchers and developers, so far, there are still very limited practical applications. The main factors for limiting oxyfuel combustion in practical engines are oxygen cost, CCS cost, necessary space for installing exhaust treatment equipment and storing CO_2 onboard.

With possible technology development and increased requirement on zero carbon emissions with low cost, hopefully oxyfuel combustion engine will be prototyped and manufactured in the near future.

Author details

Jun Peng[1*] and Xiang Li[2]

1 University of Lincoln, Lincoln, UK

2 University of Bedfordshire, Luton, UK

*Address all correspondence to: jpeng@lincoln.ac.uk

References

[1] Li X, Peng Z, Pei Y, Ajmal T, Rana K, Aitouche A, et al. Oxy-fuel combustion for carbon capture and storage in internal combustion engines—A review. International Journal of Energy Research. 2021;**2021**:1-18

[2] Serrano JR, Arnau FJ, Garcia-Cuevas LM, Farias VH. Oxy-fuel combustion feasibility of compression ignition engines using oxygen separation membranes for enabling carbon dioxide capture. Energy Conversion and Management. 2021;**247**:114732

[3] Wu Z, Wu J, Kang Z, Deng J, Hu Z, Li L. A review of water-steam-assist technology in modern internal combustion engines. Energy Reports. 2021;**7**:5100-5118

[4] Zhong S, Zhang F, Peng Z, Bai F, Du Q. Roles of CO_2 and H_2O in premixed turbulent oxy-fuel combustion. Fuel. 2018;**234**:1044-1054

[5] Kutne M, Kapadia J. Combustion instabilities in sudden expansion oxyfuel flames. Combusion and Flame. 2006; **146**(3):493-512

[6] Wu J, Yu X, Fu L, Deng J, Hu Z, Li L. A high efficiency oxyfuel internal combustion engine cycle with water direct injection for waste heat recovery. Energy. 2014;**70**:110-120

[7] Wu J, Zhang Z, Kang Z, Deng J, Li L, Wu Z. An assessment methodology for fuel/water consumption co-optimization of a gasoline engine with port water injection. Applied Energy. 2022;**310**: 118567

[8] Fu L, Wu Z, Yu X, Deng J, Hu Z, Li L. Experimental investigation of combustion and emission characteristics for internal combustion rankine cycle engine under different water injection laws. Energy Procedia. 2015;**66**:89-92

[9] Mobasheri R, Aitouche A, Peng Z, Li X. Influence of oxy-fuel combustion on engine operating conditions and combustion characteristics in a high speed direct injection (HSDI) diesel engine under homogenous charge compression ignition (HCCI) mode. SAE. 2020;**2020**:1138

[10] Mobasheri R, Aitouche A, Li X, Peng Z. Analysis of the influence of inlet temperature on oxy-fuel combustion in an hsdi diesel engine. SAE. 2022;**2022**:3

[11] Kang Z, Feng S, Lv Y, Wu J, Wu Z. Effect of direct water injection temperature on combustion process and thermal efficiency within compression ignition internal combustion Rankine engine. Case Studies in Thermal Engineering. 2021;**28**:101592

[12] European Commission. Guidelines related to the pressure equipment directive 97/23/EC. 2015. Available from: http://ec.europa.eu/DocsRoom/documents/10537

[13] Satyendra. Cryogenic process of air separation. 2013. Available from: https://www.ispatguru.com/cryogenic-process-of-air-separation/. [Accessed: July 20, 2022]

[14] Rao P, Muller MR. Industrial Oxygen: Its Generation and Use, 2007 ACEEE Summer Study on Energy Efficiency in Industry, the American Council for an Energy-Efficient Economy (ACEEE). 2007. Available from: https://www.aceee.org/files/proceedings/2007/data/papers/78_6_080.pdf. [Accessed: September 21, 2022]

[15] Sircar S, Kratz WC. Oxygen production by pressure swing

absorption. Separation Science and Technology. 1989;**24**:429-440

[16] Li X, Pei Y, Ajmal T, Rana K, Aitouche A, Mobasheri R, et al. Numerical investigation on implementing Oxy-Fuel Combustion (OFC) in an ethanol-gasoline Dual-Fuel Spark Ignition (DFSI) engine. Fuel. 2021; **302**:121162

[17] Li X, Pei Y, Li D, Ajmal T, Aitouche A, Mobasheri R, et al. Implementation of oxy-fuel combustion (OFC) technology in a gasoline direct injection (GDI) engine fueled with gasoline–ethanol blends. ACS Omega. 2021;**44**:29394-29402

Chapter 8

Recent Advances in Biodiesel from Plants

Ejiroghene Thelma Akhihiero

Abstract

Due to population explosion, and increased industrialization with urban and rural development, the need for increased energy utilization has become more intense. Petrodiesel that has been the main energy source for heavy-duty automobiles or machines has contributed immensely to environmental pollution leading to climate change, an increase in illnesses, and reduced lifespan. To combat this ugly situation arising from the utilization of Petrodiesel, biodiesel is produced from plant oil or animal fats to substitute for Petrodiesel in internal combustion engines, either as neat biodiesel or as a blend with Petrodiesel. Different scientists and researchers have produced biodiesel from edible and non-edible plant oils. Their reports show that biodiesel properties depend on the nature of the parent plant oil and the production procedures taken. These properties that are due mainly to its production procedures determining their performance in internal combustion engines. In this chapter, recent findings on biodiesel properties with their effects on performance in internal combustion engines are reviewed. Researchers' reports show that the most suitable blend of biodiesel with Petrodiesel is B20. This blend consisting of 20% biodiesel with 80% Petrodiesel has equivalent performance as Petrodiesel with fewer pollutants and only 1–7% nitrogen oxide emission.

Keywords: biodiesel, petrodiesel, compression ignition engine, non-edible oils, blending

1. Introduction

Biofuels are fuels that are produced from biomass in a short period. They are different from fossil fuels which are produced very slowly from biomass by natural processes [1]. Examples of fossil fuels are coal, crude oil or petroleum, and natural gas [2]. Examples of biofuel are fuel briquettes, bioethanol, wood, biodiesel, biobutanol, and biogas [3]. Refining of crude oil or petroleum produces fossil or conventional diesel which is the main fuel for compression ignition engine. The utilization of Petrodiesel in compression ignition engines has led to the release of greenhouse gases and particulate matter [3–10]. The release of greenhouse gases into the atmosphere is the major cause of global warming, leading to climate change and so many environmental degradations. One biofuel which has captured the interest of so many scientists and researchers is biodiesel. Biodiesel is the alkyl ester of vegetable oil or animal fats produced by transesterification under suitable reaction conditions [8, 9, 11–13].

Biodiesel sometimes called the new age fuel can proffer solutions to our global and modern energy needs either 100% or partly with Petrodiesel as a blend [9, 14–17]. Currently, biodiesel makes an insignificant contribution to the total global fuel utilization and combustion processes [18, 19]. Petrodiesel accounts for over 80% of fuel used in almost all compression ignition engines. This situation is presently unfavorable because the combustion of Petrodiesel in compression ignition engines contributes immensely to the release of greenhouse gases and severe pollution leading to global warming and climate change [6, 20, 21]. The effects of global warming and climate change are higher temperatures, rising seas, drought, floods, stronger storms, food spoilage, more heat-related illnesses such as typhoid fever, malaria fever, meningitis, etc., and economic losses [6, 22, 23].

Biodiesel burns more smoothly in compression ignition engines and produces cleaner emissions [14–16, 24–26]. Although biodiesel cannot entirely replace petroleum-based diesel fuel, it decreases a country's dependence on petroleum. Researcher's reports show that a blend of 20% biodiesel with 80% Petrodiesel in a compression ignition engine produces far less carbon IV oxide fumes with little or no greenhouse gases as compared to 100% usage of Petrodiesel which releases large emissions of carbon IV oxide fumes and other greenhouse gases [9, 21, 24, 27–32]. Total dependence on biodiesel or its blends with Petrodiesel in compression ignition engines is the sure way out of global warming and severe climate conditions experienced today [1, 6, 33–35].

Several researchers [17, 19, 21, 27–32, 36–59] have reported that the nature of plant oil together with the production processes taken affects the properties of biodiesel and its performance in compression ignition engines. Even the corrosive effects of biodiesel on aluminum, one of the metals used for engine parts were reported [60–62]. Biodiesel causes more severe pitting corrosion on aluminum than Petrodiesel [63–67]. Over sixty researchers some of whom include [17, 19, 23, 32, 36, 47–57, 61, 62, 64–70] studied alkali catalyst transesterification of different plant oil concerning their fuel properties such as density, viscosity, cetane number, cold flow properties, flash point, and calorific value. Their studies show the best quality alkyl ester under different transesterification circumstances [1, 14, 17, 19, 24, 35, 36, 47–57, 71–73]. A review on biodiesel fuel specifications by [61] found that over 80% of scientists and researchers failed to extensively report on the fuel properties of biodiesel and their influence on performance in compression ignition engines.

Among the plant oil biodiesel studied, *Jatropha curcas* biodiesel and algae biodiesel stands out in terms of quality and good fuel properties for use in compression ignition engines [36, 47, 48, 56, 71, 74, 75]. Some studies [5, 25, 56, 76–78] even report that *Jatropha* seed oil or its blend with conventional diesel fuel can be used as diesel in compression ignition engines without engine modification.

Jatropha is a non-edible plant, hence its usage in fuel production does not compete with food. Moreover, *J. curcas* has a yield per hectare of more than four times that of soybean and ten times that of corn [79]. Recent studies on the evaluation of 75 non-edible oils as biodiesel feedstocks, identified *J. curcas* as one of the most promising plants apart from algae [71, 79, 80]. *J. curcas* plant is very prolific, it grows on any type of soil, gets matured within a very short period, and produces large quantities of seed almost throughout the year. The percentage oil content in *J. curcas* seed is between 28 and 60% [79–81]. This chapter reports the viability and sustainability of *J. curcas* biodiesel and algae biodiesel over several plant biodiesel. The biodiesel fuel properties of some common plant oils in comparison to those of *J. curcas* biodiesel and algae biodiesel are reviewed.

2. Biodiesel feedstocks from plants

There are several ways to obtain biodiesel feedstocks from plants. They include extraction of the plant oil by solvent extraction or by hydraulic press, supercritical fluid extraction, ultrasound-assisted extraction, and microwave-assisted extraction [61, 74, 81]. These methods are suitable for different plant materials, depending on the nature of the plant material and the desired yield [1, 71, 75, 81].

Several oils extracted from plants including algae oil have been used as biodiesel feedstocks [24, 71, 79]. To assess the quality of oil for biodiesel production, the fatty acid composition and free fatty acid content of the oil must be determined. Oils with high free fatty acids of over 3% must be esterified to reduce their free fatty acids before transesterification to biodiesel [74, 76, 77].

The fatty acid composition of the plant oil is an indication of its biodiesel quality and performance in compression ignition engines especially its cold flow properties [8, 71, 74–76]. Plant oils whose structural fatty acid content has a sufficient number of double carbon bonds show promising biodiesel qualities such as good cetane number, good oxidative stability, lower viscosity with good engine atomization, and higher calorific value [1, 71, 72, 74]. **Table 1** shows some common biodiesel feedstocks with their fatty acid content.

3. Biodiesel processing

According to Rudolf Diesel [83], pure vegetable oil could be used as a biofuel but for its high viscosity which causes lots of problems to internal combustion engines. Some researchers have proven through research that by blending a suitable amount of pure vegetable oil with a suitable amount of Petrodiesel, one could overcome the challenges of using pure vegetable oil in internal combustion engines [15, 16, 24, 26]. This was proven to be true for pure *J. curcas* seed oil. The fuel produced by the blending of a suitable amount of *J. curcas* seed oil with Petrodiesel possesses the ideal quality of fuel that burns smoothly in an internal combustion engine with little or no emission of dangerous fumes [56, 74]. For vegetable oil to possess properties similar to that of Petrodiesel and to burn smoothly in internal combustion engines, it has to be modified by transesterification [4, 74, 75, 84].

Transesterification is a chemical reaction between the high molecular weight triglyceride in vegetable oil or animal fat with low molecular weight or short-chain alcohol such as methanol or ethanol to produce methyl or ethyl esters and a glycerol molecule. Transesterification is an equilibrium reaction that takes place in three consecutive steps. One of the reasons for determining the fatty acid contents of oil is to enable one to calculate the oil's average molecular weight. Knowing the average molecular weight of an oil is an important step in knowing the required amount of alcohol needed for transesterification reaction for a given oil to alcohol molar ratio. For example, the average molecular weight of *Jatropha* oil is 900 moles, while that of algae oil is 274.4 moles [11, 12]. Certain variables are known to affect the transesterification reaction of vegetable oil or animal fats. These variables include temperature, catalyst type, catalyst amount, the molar ratio of oil to alcohol, reaction time, stirring

Oil or fat	$C_{14:0}$ Myristic acid	$C_{16:0}$ Palmitic acid	$C_{16:1}$ Palmitoleic acid	$C_{18:0}$ Stearic acid	$C_{18:1}$ Cis Oleic acid	$C_{18:1}$ trans oleic acid	$C_{18:2}$ Linoleic acid	$C_{18:3}$ Linoleic acid	$C_{20:0}$ Arachidic acid	$C_{22:0}$ Erucic acid
Soybean		6–10		2–5	20–30		50–60	5–11		
Corn	1–2	8–12		2–5	19–49		34–62	Trace		
Peanut		8–9		2–3	50–65		20–30			
Olive		9–10		2–3	73–84		10–12	Trace		
Safflower		4.8		1.4	74.1		19.7			
Rapeseed		3.0		0.8	13.1		14.1	9.7	7.4	50.7
Lard	1–2	28–30		12–18	40–50		7–13	0.1		
Tallow	3–6	24–32		20–25	37–43		2–3			
Yellow grease	1.27	17.44		12.38	54.67		7.96	0.69	0.25	0.52
Linseed oil		4–7		2–4	25–40		35–40	25–60		
Jatropha	0.26	14.398	0.704	6.65	43.39		32.98	0.351	0.218	0.506
Algae	1.0	16.6	1.9	1.9	26.9	3.7	22.8	16.1	1.5	
Cotton seed	0–2	20–25		1–2	23–35		40–50	Trace		
Sunflower		6.6	0.1	3.08	17.31	73.31				

Table 1.
Biodiesel Feedstocks with their fatty acid composition [1, 71, 72, 79, 81, 82].

speed, type of oil, free fatty acid of oil or fats, and so on. Many researchers have reported how each of these variables affects the transesterification reaction of vegetable oil or animal fats to biodiesel. Their reports also include the effect of these variables on the quality of biodiesel produced and how these variables can be modified to give higher yield and best quality biodiesel [21, 40, 53, 77, 85]. Of these variables oil to alcohol molar ratio, temperature, catalyst type, and catalyst concentration or catalyst amount play vital roles [12, 77, 82]. To cause the reaction to proceed at a reasonable speed, a catalyst must be involved. There are three main types of catalyst, namely liquid catalyst examples are a solution of sodium hydroxide or potassium hydroxide, solid catalyst examples are calcium oxide, zinc oxide, zeolite, etc. and enzyme catalyst example is lipase. Researchers have observed that liquid catalysts cause transesterification reactions to proceed fasters and products are formed in a shorter time than solid and enzyme catalysts [4, 12, 82].

Figure 1 shows the biodiesel chemical reaction while Eq. (1)–(23) shows the chemical reaction mechanism.

The biodiesel or transesterification reaction is of the form [3, 50].

$$A + B \overset{Catalyst}{\rightleftharpoons} C + D \tag{1}$$

Where A = Triglyceride (TG), B = alcohol (ROH).
C = Glyceride and D = methyl esters (ME).
Therefore Eq. (1) becomes

$$TG + ROH \overset{Catalyst}{\rightleftharpoons} Glyceride + ME \tag{2}$$

Where the glyceride could be diglyceride, monoglyceride, or glycerol.

Eq. (2) [3, 12] is an equilibrium reaction that occurs in three consecutive steps forming intermediates. Stoichiometrically, 3 moles of alcohol were required to react with 1 mole of TG to form 3 moles of esters and 1 mole of Glycerol [12, 77, 82]. Therefore, we have [3],

$$TG + ROH \underset{K_2}{\overset{K_1}{\rightleftharpoons}} DG + ME \tag{3}$$

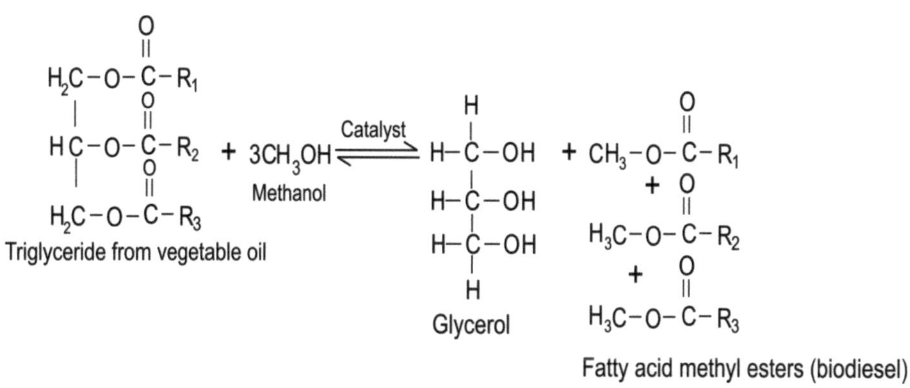

Figure 1.
Transesterification of triglycerides with methanol to form fatty acid methyl esters [77].

$$DG + ROH \underset{K_4}{\overset{K_3}{\rightleftharpoons}} MG + ME \tag{4}$$

$$MG + ROH \underset{K_6}{\overset{K_5}{\rightleftharpoons}} GL + ME \tag{5}$$

The overall reaction is, therefore [12, 82].

$$TG + 3ROH \rightleftharpoons GL + 3ME \tag{6}$$

From Eq. (3).
$TG + ROH \overset{K_1}{\rightarrow} DG + ME$ (7)
and

$$DG + ME \overset{K_2}{\rightarrow} TG + ROH \tag{7}$$

From Eq. (4).
$DG + ROH \overset{K_3}{\rightarrow} MG + ME$ (9)
and

$$MG + ME \overset{K_4}{\rightarrow} DG + ROH \tag{8}$$

from Eq. (5),
$MG + ROH \overset{K_5}{\rightarrow} GL + ME$ (11)
and

$$GL + ME \overset{K_6}{\rightarrow} MG + ROH \tag{9}$$

The reactions above are initiated and terminated as follows:
Formation of alkoxide from methanol and sodium hydroxide [12].

$$\underset{alcohol}{ROH} + \underset{\substack{from\\catalyst}}{OH^-} \rightleftharpoons \underset{alkoxide}{RO^-} + \underset{water}{H_2O} \tag{10}$$

The alkoxide ion attacks the glyceride molecules to form methyl esters and glycerol as shown in Eqs. (14)–(19).
Transesterification [12, 82]:

$$TG + RO^- \rightleftharpoons DG^- + ME \tag{11}$$

$$DG^- + ROH \rightleftharpoons DG + RO^- \tag{12}$$

$$DG + RO^- \rightleftharpoons MG^- + ME \tag{13}$$

$$MG^- + ROH \rightarrow MG + RO^- \tag{14}$$

$$MG + RO^- \rightarrow GL^- + ME \tag{15}$$

$$GL^- + ROH \rightarrow GL + RO^- \tag{16}$$

If saponification, an almost inevitable side reaction associated with transesterification when using a homogeneous catalyst such as sodium hydroxide is allowed to take place, the following reaction equations Eqs. (20)–(23) would be observed [12, 76].

$$ME + OH^- \rightarrow ROH + Soap \quad (17)$$

$$TG + OH^- \rightarrow DG + Soap \quad (18)$$

$$DG + OH^- \rightarrow MG + Soap \quad (19)$$

$$MG + OH^- \rightarrow GL + Soap \quad (20)$$

Saponification reactions of Eqs. (20)–(23) become prominent during transesterification when using vegetable oil with high free fatty acids and high moisture content with a homogenous catalyst like sodium hydroxide [76]. Therefore, transesterification of high-free fatty acid vegetable oil with sodium hydroxide catalyst would result in a lot of soap formation with little or no biodiesel [4, 76, 77]. Hence it is recommended that if sodium hydroxide catalyst must be used in the transesterification of oil with high free fatty acid the oil must be esterified to a free fatty acid of below 3% and the oil must be dried or free from moisture [12, 76, 77].

The biodiesel produced by transesterification from *Jatropha curcas* seed oil with methanol using sodium hydroxide as a catalyst has properties close to those of conventional diesel fuel [3, 8]. It was suitable to be used as neat biodiesel in internal combustion engines. Its blend with conventional or Petrodiesel also could result in low carbon emissions during combustion. Biodiesel produced from vegetable oils or animal fats with a catalyst concentration of 1%, the molar ratio of oil to alcohol of 6:1 or 8:1, and at a temperature of 60°C or 65°C gave a good quality fuel [3, 7–9].

Biodiesel produced by this method has properties close to that of Petrodiesel. Its viscosity becomes lower than that of pure vegetable oil to enable it to burn smoothly in an internal combustion engine. It has a high cetane number and higher flash point than Petrodiesel which makes it safer to transport than Petrodiesel [8, 71, 74, 75].

Table 2 compares the properties of biodiesel produced by various researchers with those of conventional diesel fuel and ASTM biodiesel standards.

4. Biodiesel fuel properties

The fuel properties of biodiesel and its performance in a compression ignition engine can be predicted mainly from the type of fatty acids contained in it. Plant oils or fats which contain a substantial number of double carbon bonds in their structure are a sure indication of producing quality biodiesel [36, 71, 72, 79, 86]. Biodiesel properties such as cetane number, calorific value, viscosity, oxidative stability, flash point, pour point, and cloud point are connected with the type and nature of the carbon structure of the parent oil. The free fatty acid of the parent oil is also a factor capable of affecting the biodiesel fuel property and production cost [71, 76, 87]. Oils or fats with high free fatty acid usually must go through a two-step transesterification process with a homogenous catalyst [8, 76]. The additional process of esterification of high free fatty acid oils to reduce the free fatty acids before trans esterifying it to biodiesel affects its properties and increases the production cost [11]. Some *Jatropha* oil and algae oil have low free fatty acid making them suitable for a one-step transesterification to biodiesel [36, 71, 74, 75]. **Table 3** shows the reviewed biodiesel fuel properties of some common feedstocks.

Parameters	*Jatropha* Biodiesel @ 65°C, 1% NaOH, Molar ratio of methanol to oil 8:1 [76]	*Jatropha* Biodiesel @ 65°C, 1% CaO, Molar ratio of methanol to oil 8:1 [76]	Palm Kernel Oil Biodiesel [8]	B20 Biodiesel blend of 20% biodiesel with 80% fossil diesel [8]	Conventional Diesel ASTM D975 [9]	Biodiesel Fuel Properties (ASTM D6751) [9, 61]
Flash point, °C	191	192	171	66	60–80	100–170
Cloud point, °C	7	9	9.9	6	−15 to 5	−3 to 12
Pour point, °C	11	12	−1.0	1	−35 to −15	−15 to 10
Kinematic Viscosity @ 40°C, mm²/s	4.91	4.86	4.87	8.79	4.0–6.0	1.3–4.1
Calorific value (MJ/Kg)	42.56	42.22			45.5	39-43
Cetane number	49.17	48.94			48–65	40–55
Specific gravity	0.891	0.893	0.898	0.832	0.88	0.85
Acid number (mgKOH/g)	0.45	0.47				

Table 2.
Biodiesel properties of some vegetable oils compared with standards and conventional diesel properties.

5. Biodiesel tests on engines

Many scientists and researchers [1, 8, 9, 61, 71, 75, 76, 88, 89] have assessed the performance of biodiesel produced on compression ignition engines. Their reports show that once the biodiesel fuel properties fall within the standard ASTM D6751 and EN14214 diesel fuel properties, the biodiesel is sure to have good performance in any compression-ignition engines.

The work of over 60 biodiesel researchers on biodiesel fuel quality and performance on compression ignition engines was reviewed [1, 14–16, 20, 24, 71, 72]. Their reports confirmed that all produced biodiesel with fuel properties conforming to ASTM D6751 or EN14214 have significant superiority to conventional diesel fuel due to total hydrocarbon volume and other vital fuel properties [7, 71, 77, 88]. These biodiesels have better performance in compression ignition engines with very low carbon monoxide and carbon dioxide emission, low particle substance, and ultra-low Sulfur content [1, 71, 72]. The cold flow properties of biodiesel were also found to be very encouraging, with that of algae and *Jatropha* biodiesel being the best amongst all others [71, 76].

The kinetic viscosity of fuel determines its flow, spray, and atomization properties. High density creates high viscosity leading to inefficient combustion, poor engine performance, and excess carbon monoxide emission [71, 72, 74]. The density of algae and *Jatropha* biodiesel from several research show conformities to standards [71, 74, 76].

The performance of a blend of biodiesel with conventional diesel fuel was also examined. The findings show that an increase in the proportion of biodiesel in the blend decreases carbon monoxide and carbon dioxide emissions, but also increases

Properties/Units	*Jatropha* Biodiesel [56, 61, 74, 76]	Sunflower Biodiesel [7, 61]	Algae Biodiesel [61, 71, 87]	Pongamia Pinnata Biodiesel [34, 61, 72]	Almond Biodiesel [9, 61]
Flash point °C	148–170	110–183	150	90–172	172
Pour point °C	−12 to 2.0		−2.67		
Cloud point °C	4–10.2		1		
Kinematic Viscosity @ 40°C (mm^2/s)	4.23–5.65	4.10–4.90	3.16	4.16–10.64	4.2
Density (g/cm^3)	0.870–0.880	0.860–0.885	0.880	0.878–0.920	0.876–0.891
Cetane number	4–57.4		47	53–55.7	
Calorific value (MJ/kg)	38.5–45.5	38.47–45.3	40.66	35.560	
Oil content (in %)	28–60	40		30–70	
Acid number (mgKOH/g)				0.497	

Table 3.
Reviewed biodiesel fuel properties for some common feedstocks.

nitrogen oxide emissions which can be reduced with ionic liquid additives [62]. The emission of oxides of nitrogen is increased with an increase of biodiesel in the blend, especially algae biodiesel [15, 26]. This is not surprising because algae are known to assimilate nitrogen from their marine environment [2]. Biodiesel is known to increase the corrosion of vessels because of its high oxygen content, especially pitting corrosion [63].

Soladiesel BDR company is among other companies reported to have used different biodiesels with standard diesel engines without modification. The company reported that biodiesel emits fewer pollutants than conventional diesel and has equivalent performance to Petrodiesel with lesser engine wear [71, 74, 75, 87].

Several automobile manufacturers including Caterpillar, DEUTZ, MAN, and Volvo have built trucks, buses, and engines that can run on pure and blended biodiesel [90, 91]. A report shows that using biodiesel in heavy-duty vehicles reduces greenhouse gases emission by up to 90% [91]. Biodiesel raises the cetane number of blended fuel and improves fuel lubricity [67, 92] which is usually low in conventional low sulfur diesel [93]. A higher cetane number makes the engine easier to start and reduces ignition delay, together with improved lubricity, premature engine wear is prevented. Hence, the blending of biodiesel with Petrodiesel in internal combustion engines is highly recommended.

6. Economics of biodiesel production

A major hurdle facing commercial biodiesel production is the cost of fats and oils. About 60–80 percent of the cost of producing biodiesel arises from oil seed procurement, transport, seed storage, and oil extraction [11]. The cost depends mostly on the type and nature of feedstock used.

The use of lower-cost feedstock such as waste cooking oil, tallow, and non-edible oils would reduce the production cost of biodiesel and biodiesel blend fuel. A blended fuel of 20 percent biodiesel and 80 percent Petrodiesel would reduce cost. For bioethanol production, the use of sugarcane bagasse, cassava peels instead of tubers, corn cobs, or other non-edible cellulosic biomass helps to reduce production costs [84].

The cost of biodiesel can also be reduced by the use of an appropriate catalyst. Biodiesel produced with liquid catalysts usually contain impurities, excess alcohol, and parts of the catalysts used in the process [4]. The excess alcohol can be removed by using a flash evaporation process or by distillation. The biodiesel fuel after being separated from glycerin is further purified by washing the fuel with hot or warm water to remove unreacted alcohol, alkaline, oil, or glycerol/glycerides. The cost of this additional refining process adds to the already high manufacturing cost of biodiesel due to the high cost of feedstock [4, 75]. Akhihiero [11] calculated the approximate cost of production of a liter of *Jatropha* biodiesel to be $9.01 (₦3740.50). This cost which was obtained in 2019, could be reduced by adopting a better production process such as the continuous process instead of the batch process [12]. The cost can also be minimized by using the fuel as a blend with Petrodiesel. Research shows that a fuel produced with 20% biodiesel with 80% Petrodiesel performs excellently well in internal combustion engines by emitting low carbon deposits with smooth burning and complete combustion [15, 87].

Despite the high production cost of biodiesel by transesterification, its discovery and production are a welcome development. The environmental damage caused by the complete utilization of Petrodiesel or fossil fuels together with their effects on human health far outweighs the high cost of biofuels. A compromise is reached by blending which reduces cost drastically while still overcoming the negative effect of fossil utilization [9, 14, 16].

7. Advantages and disadvantages of biofuels

Benefits of biofuel usage include fewer carbon emissions, when burned they release as much carbon as they absorbed during the growth of plants. This is not the case with fossil fuels which release excess carbon and oxides of carbon.

Other benefits include energy efficiency, reducing petroleum oil dependency, a healthy environment, positive economic impact, reducing greenhouse gases and reducing global warming, sustainability, high-quality engine performance, and lowering particulate matters [1, 20, 71, 72].

Disadvantages of biofuel include the need for a lot of labor, the high cost of production, and various health problems that may be encountered during processing. Cases of heart disease, respiratory symptoms such as asthma, chronic bronchitis, or even premature death may result from biofuel processing [94]. Sodium methoxide the popular liquid catalyst for biodiesel production is very toxic to humans. To avoid the menace of sodium methoxide, solid catalysts such as calcium oxide, zinc oxide, doped or modified eggshells, or animal bones are used in biodiesel production [8, 39–43].

8. Challenges posed by biodiesel

Biodiesel utilization is not without challenges. These challenges include;

1. Nitrogen oxides emission.

2. Corrosion of aluminum

3. High production cost

4. Production health hazards

8.1 Nitrogen oxides emission

One major challenge of biodiesel utilization in compression ignition engines is the high emission of Nitrogen oxides. Nitrogen oxides are pollutants that are responsible for the corrosion of metals and acid rain [62]. Neat biodiesel is known to emit higher oxides of Nitrogen than Petrol diesel [21, 27–32]. The emission of Nitrogen oxide by biodiesel is relatively 10–30% higher than that of Petrodiesel in compression ignition engines [27–29, 68, 69]. A blend of 20% biodiesel to Petrodiesel-only emits 1–7% Nitrogen oxide [23, 62, 68, 69]. As the number of biodiesel increases in the blend, Nitrogen oxide emission increases. Nitrogen oxide emission depends on the type of biodiesel feedstock [27, 62, 68–70]. The highest Nitrogen oxide emission was reported with the most highly unsaturated biodiesel such as *Jatropha,* neem, castor, Karanja, algae, etc. [62, 68–70].

Biodiesel produced from feedstocks with a high saturation of fatty acids such as palm oil or animal fats like tallow emits fewer Nitrogen oxides in Compression ignition engines. An increase in Nitrogen oxide emission also depends on engine technology [64–67, 85, 95]. Studies reveal that Nitrogen oxide emission increases with newer engines [73, 85, 96]. However, a blend of 20% biodiesel with Petrodiesel shows little or no emission of Nitrogen oxides [17, 35, 62, 64–67, 69, 70, 73, 85, 95, 96].

8.2 Corrosion of aluminum

Biodiesel causes corrosion in aluminum, especially pitting corrosion, more than Petrodiesel. This is because its oxygen content is higher than that of Petrodiesel [63]. The corrosion of aluminum by biodiesel can be controlled with the use of plant extracts like the ethanol extract of Rosemary leaves [66] and Vitex negundo leaf extract [67] and ionic liquids as corrosion inhibitors [85, 96]. Corrosion inhibitors also help to reduce nitrogen oxide emission [73, 85, 95, 96].

8.3 High production cost

The high production cost of biodiesel has been the major reason why biodiesel utilization in compression ignition engines is unpopular in many countries, especially Nigeria [13, 97–109]. The feedstock cost is the major contributor to this high cost [11]. Hence, it's been recommended to use waste feedstocks and oils of non-edible plants [11, 53, 54, 97]. Three ways to reduce biodiesel production costs are stated thus:

I. The use of low-cost feedstocks with high oil content such as low-cost *J. curcas* and algae for biodiesel production [97, 110, 111]

II. The use of heterogeneous catalysts and plant oils with low free fatty acids [8, 107–109]. The use of heterogeneous catalysts prevents the additional cost of product washing which is associated with homogeneous catalysts. Also, oils with free fatty acids lower than 3% are suitable with homogeneous catalysts in one-step transesterification. Oils with high free fatty acid must undergo the additional cost of esterification to prevent soap formation [39–45, 76], before transesterification to biodiesel [8, 108].

III. Blending of biodiesel with Petrodiesel in a ratio of 20% biodiesel to 80% Petrodiesel. For example, from the work of [11], a 100% *Jatropha* biodiesel has a production cost of $9.01/L. By producing a 20% blend with Petrodiesel, which has a global average cost of $1.40/L [112] the estimated cost of the B20 blend is about $2.92/L.

8.4 Production health hazards

The use of acid or alkaline heterogeneous catalysts such as calcium oxide, zinc oxide, zeolite, doped aluminum oxide, doped eggshell, etc. [39–46] in transesterification eliminates the health hazard in biodiesel production [8, 35, 58, 59]. The only drawback with heterogeneous catalysis is a longer reaction time [8]. The reaction could take additional 30 minutes or over an hour to complete with heterogeneous catalysts [8, 59, 76].

In biodiesel production, the most popular initial catalyst is sodium hydroxide [4, 76]. Sodium hydroxide in methanol produces sodium methoxide which is very toxic to humans [8]. Hence, the usage of face masks is recommended during the production of biodiesel, especially with homogeneous catalysts.

9. Conclusion

Research has revealed and is still revealing what is embedded in the numerous biomass all around us. Biofuel produced from biomass has a lot to offer in combating global warming, reducing the effects of climate change, and reducing disease and environmental degradation. Sole utilization of neat biodiesel or a blend of biodiesel with Petrodiesel is recommended in all compression ignition engines to maintain a cleaner and safer environment. However, the high cost of biodiesel is a challenge that can be controlled by adopting a better production process and by using a blend of biodiesel with Petrodiesel. Biodiesel quality and performance characteristics have been traced to the chemical structure of its parent oil and production procedures adopted. Alkali catalyst transesterification of low-free fatty acid oils is one of the most promising methods to produce cheaper and better-quality biodiesel for compression ignition engines. Low free fatty acid *J. curcas* oil and Algae oils have shown promising features in biodiesel production for compression ignition engines. The performance of *Jatropha* biodiesel and algae biodiesel in compression ignition engines has been outstanding. They both have good cold flow properties because of the presence of palmitoleic acid and high oleic and linoleic acids in them. However, the increase in emission of Nitrogen oxides by biodiesel or its blend can be controlled or prevented by using plant extracts or ionic liquids as corrosion inhibitors. The optimal biodiesel blend for a more tolerable Nitrogen oxide emission for compression ignition engines is B20. Also, the use of low-cost feedstock with high oil content such as low-cost *J. curcas* and algae for biodiesel production is recommended. However, the use of corrosion inhibitors such as Rosemary leaves extract on *Jatropha* and algae biodiesel is a must due to the presence of high unsaturated fatty acids which lead to high nitrogen oxides emission.

Usage of acid or alkaline heterogeneous catalysts in transesterification of plant oils is the sure way to prevent health hazards associated with homogenous catalysis in biodiesel production. Also, proper use of safety masks during processing is another option to avoid the risk of inhaling poisonous Sodium methoxide.

The production cost of a liter of *Jatropha* biodiesel was calculated to be \$9.01/L (₦3740.50/L) but this cost can be potentially reduced to \$2.92/L by producing a B20 blend. Because the cetane number of *Jatropha* and algae biodiesel is higher, it becomes easier to blend either of them with Petrodiesel at higher concentrations.

Author details

Ejiroghene Thelma Akhihiero
University of Benin, Benin City, Nigeria

*Address all correspondence to: thelma.akhihiero@uniben.edu

References

[1] Kurniasih E, Pardi P. Analysis of process variables on biodiesel transesterification reaction using Taguchi Method. In: IOP Conference Series: Materials Science and Engineering. Vol. 420. Medan, Indonesia: 2nd Nommensen International Conference on Technology and Engineering; 2018

[2] Available from: https://www.clarku.edu/faculty/roberton. [Accessed: June 29, 2022]

[3] Akhihiero ET, Audu TOK, Aluyor EO. Kinetics and Modelling of Jatropha curcas Biodiesel Reaction at 32°C. Petroleum Technology Development Journal. 2016;**6**(1):78-90

[4] Akhihiero ET, Aluyor EO, Audu TOK. The effect of catalyst phase on biodiesel production. (A review). Journal of Engineering and Applied Sciences. 2010;**2**:93-102

[5] Akhihiero ET, Ayodele BV, Gbegudu ET. Optimization of biodiesel production using response surface methodology and central composite design. Journal of the Nigeria Association of Mathematical Physics. 2013;**23**:585-590

[6] Akhihiero ET, Odu GO. Environmental impact of climate change and the renewable energy options for sustainable development. International Journal of Innovative Environmental Studies Research. 2014;**2**(1):63-65

[7] Akhihiero ET, Omorewa YG, Ebhodaghe SO. Effect of blending ratio on the properties of sunflower biodiesel. Journal of Materials and Environmental Science. 2019;**10**(10):987-994

[8] Akhihiero TE. Solid catalyzed reaction of jatropha curcas seed oil with methanol. European Journal of Sustainable Development Research. 2020;**4**(1):em0105. DOI: 10.29333/ejosdr/6263

[9] Akhihiero ET, Ebhodaghe SO. Effect of blending ratio on the fuel properties of almond biodiesel. European Journal of Sustainable Development Research. 2020;**4**(3):em0119. DOI: 10.29333/ejosdr/7804

[10] Raqeeb MA, Bhargavi R. Biodiesel production from waste cooking oil. Journal of Chemical and Pharmaceutical Research. 2015;**7**(12):670-681

[11] Akhihiero ET, Ebhodaghe SO. Feasibility of Biodiesel Production from Jatropha curcas Oil. Journal of Civil and Environmental Systems Engineering. 2020;**18**(1)

[12] Akhihiero ET, Ayodele BV, Alsaffar MA, Audu TOK, Aluyor EO. Kinetic studies of biodiesel production from jatropha curcas oil. Journal of Engineering. 2021;**27**:33-45. DOI: 10.31026/j.eng.2021.04.03

[13] Mishra VK, Goswami R. A review of production, properties and advantages of biodiesel. Biofuels. 2018;**9**(2):273-289

[14] Ilkilic C. The effect of sunflower oil methyl Ester and Diesel fuel blend on the performance of a diesel engine. Energy Sources, Part A: Recovery, Utilization, and Environmental Effects. 2008;**30**:1761-1770. DOI: 10.1080\15567030701268443

[15] Igbokwe JO, Nwufo OC, Nwaiwu CF. Effects of blends on the properties, performance, and emission of palm kernel oil biodiesel. Biofuels. 2015;**6**(1):1-8. DOI: 10.1080/17597269.2015.1030719

[16] Khan K, Kumar G, Sharma AK, Kumar PS, Mandal C, Chintala V. Performance and emission characteristics of a diesel engine using a complementary blending of castor and Karanja biodiesel. Biofuels. 2016;**9**(1):53-60. DOI: 10.1080/17597269.2016.1256552

[17] Sani S, Kaisan MU, Kulla DM, Obi AI, Jibrin A, Ashok B. Determination of physico chemical properties of biodiesel from Citrullus lanatus seeds oil and diesel blends. Industrial Crops and Products. 2018;**122**:702-708

[18] Available from: https://www.conserve.energy.future.com. [Accessed: June 15, 2022]

[19] Sierra-Cantor JF, Guerrero-Fajardo CA. Methods for improving the cold flow properties of biodiesel with high saturated fatty acids content: A review. Renewable and Sustainable Energy Reviews. 2017;**72**:774-790

[20] Battaglia SM. Shifting weather patterns in the warming Arctic: The scandes case. Weatherwise. 2018;**72**(1): 23-29. DOI: 10.1080/00431672.2019.153

[21] Sinha D, Murugavelh S. Biodiesel production from waste cotton seed oil using low cost catalyst: Engine performance and emission characteristics. Perspectives in Science. 2016;**8**:237-240

[22] Rodrigues-Ake A, Xaman J, Hernandez-Lopez I, Sauceda D, Carranza-Chavez FJ, Zavala-Gllen I. Numerical study and thermal evaluation of a triple glass window under Mexican warm climate conditions. Elsevier Journal of Energy. 2022;**239**(Part B): 122075

[23] Prabu A, Anand RB. Emission control strategy by adding alumina and cerium oxide nano particle in biodiesel. Journal of the Energy Institute. 2016;**89**(3):366-372

[24] Das D, Pathak V, Yadav AS, Uphadyaya R. Evaluation of performance, emission, and combustion characteristics of a diesel engine fueled with castor biodiesel. Biofuels. 2017;**8**(2): 225-233. DOI: 10.1080/17597269.2016.1221298

[25] El-Diwani G, Attai NK, Hawash SI. Development and evaluation of biodiesel fuel and byproducts from Jatropha oil. International journal of Environmental Science and Technology. 2009;**6**(2): 219-224

[26] Catapano F, Dilorio S, Magno A, Vaglieco BM. Effect of fuel quality on combustion evolution and particle emissions from PFI and GDI engines fueled with gasoline, ethanol, and blend, with a focus on 10–23nm particles. Elsevier Journal of Energy. 2022;**239**(Part B):122198

[27] Roy MM, Wang W, Bujold J. Biodiesel production and comparison of emissions of a DI diesel engine fueled by biodiesel–diesel and canola oil–diesel blends at high idling operations. Applied Energy. 2013;**106**:198-208

[28] Davidson EA, Kanter D. Inventories and scenarios of nitrous oxide emissions. Environmental Research Letters. 2014;**9**(10):105012

[29] Chang YC, Lee WJ, Wu TS, Wu CY, Chen SJ. Use of water containing acetone–butanol–ethanol for NOx-PM (nitrogen oxide-particulate matter) trade-off in the diesel engine fueled with biodiesel. Energy. 2014;**64**:678-687

[30] Roy MM, Wang W, Alawi M. Performance and emissions of a diesel engine fueled by biodiesel–diesel, biodiesel–diesel-additive and kerosene–

biodiesel blends. Energy Conversion and Management. 2014;**84**:164-173

[31] Mirzajanzadeh M, Tabatabaei M, Ardjmand M, Rashidi A, Ghobadian B, Barkhi M, et al. A novel soluble nano-catalysts in diesel–biodiesel fuel blends to improve diesel engines performance and reduce exhaust emissions. Fuel. 2015;**139**:374-382

[32] Soudagar MEM, Nik-Ghazali NN, Kalam MA, Badruddin IA, Banapurmath NR, Akram N. The effect of nano-additives in diesel-biodiesel fuel blends: A comprehensive review on stability, engine performance and emission characteristics. Energy Conversion and Management. 2018;**178**:146-177

[33] Borugadda VB, Goud VV. Biodiesel production from renewable feedstocks: Status and opportunities. Journal of Renewable and Sustainable Energy Reviews. 2012;**16**(7):4763-4784

[34] Ahmad M, Zafar M, Khan MA, Sultana S. Biodiesel from Pongamia pinnata L. oil: A promising alternative bioenergy source. Energy Sources, Part A: Recovery, Utilization, and Environmental Effects. 2009;**31**:1436-1442

[35] Serin H, Ozcanli M, Kemal Gokce M, Tuccar G. Biodiesel production from tea seed (camellia sinensis) oil and its blends with diesel fuel. International Journal of Green Energy. 2013;**10**(4):370-377

[36] Islam MA, Magnusson M, Brown RJ, Ayoko GA, Nurun Nabi M, Heimann K. Microalgal species selection for biodiesel production is based on fuel properties derived from fatty acid profiles. Energies. 2013;**6**:5676-5702. DOI: 10.3390/en6115676

[37] Rizwanul Fattah IM, Ong HC, Mahlia TMI, Mofijur M, Silitonga AS, Rahman SA, et al. State of the art of catalysts for biodiesel production. Frontiers in Energy Research. 2020;**8**:101

[38] Di Serio M, Tesser R, Pengmei L, Santacesaria E. Heterogeneous catalysts for biodiesel production. Energy & Fuels. 2008;**22**(1):207-217

[39] Islam A, Taufiq-Yap YH, Chu CM, Chan ES, Ravindra P. Studies on design of heterogeneous catalysts for biodiesel production. Process Safety and Environmental Protection. 2013;**91**(1–2): 131-144

[40] Marinković DM, Stanković MV, Veličković AV, Avramović JM, Miladinović MR, Stamenković OO, et al. Calcium oxide as a promising heterogeneous catalyst for biodiesel production: Current state and perspectives. Renewable and Sustainable Energy Reviews. 2016;**56**:1387-1408

[41] Faruque MO, Razzak SA, Hossain MM. Application of heterogeneous catalysts for biodiesel production from microalgal oil - A review. Catalysts. 2020;**10**(9):1025

[42] Marwaha A, Dhir A, Mahla SK, Mohapatra SK. An overview of solid base heterogeneous catalysts for biodiesel production. Catalysis Reviews. 2018;**60**(4):594-628

[43] Atadashi IM, Aroua MK, Aziz AA, Sulaiman NMN. The effects of catalysts in biodiesel production: A review. Journal of Industrial and Engineering Chemistry. 2013;**19**(1):14-26

[44] Balajii M, Niju S. Biochar-derived heterogeneous catalysts for biodiesel production. Environmental Chemistry Letters. 2019;**17**(4):1447-1469

[45] Buasri A, Chaiyut N, Loryuenyong V, Wongweang C, Khamsrisuk S.

Application of eggshell wastes as a heterogeneous catalyst for biodiesel production. Sustainable Energy. 2013;**1**(2):7-13

[46] Suryaputra W, Winata I, Indraswati N, Ismadji S. Waste capiz (Amusium cristatum) shell as a new heterogeneous catalyst for biodiesel production. Renewable Energy. 2013;**50**:795-799

[47] Patel A, Arora N, Mehtani J, Pruthi V, Pruthi PA. Assessment of fuel properties on the basis of fatty acid profiles of oleaginous yeast for potential biodiesel production. Renewable and Sustainable Energy Reviews. 2017;**77**:604-616

[48] Folayan AJ, Anawe PAL, Aladejare AE, Ayeni AO. Experimental investigation of the effect of fatty acids configuration, chain length, branching and degree of unsaturation on biodiesel fuel properties obtained from lauric oils, high-oleic and high-linoleic vegetable oil biomass. Energy Reports. 2019;**5**:793-806

[49] Rajagopal K, Bindu C, Prasad RBN, Ahmad A. The effect of fatty acid profiles of biodiesel on key fuel properties of some biodiesels and blends. Energy Sources, Part A: Recovery, Utilization, and Environmental Effects. 2016;**38**(11):1582-1590

[50] Gopinath A, Sairam K, Velraj R, Kumaresan G. Effects of the properties and the structural configurations of fatty acid methyl esters on the properties of biodiesel fuel: A review. Proceedings of the Institution of Mechanical Engineers, Part D: Journal of Automobile Engineering. 2015;**229**(3):357-390

[51] Islam MA, Brown RJ, Brooks PR, Jahirul MI, Bockhorn H, Heimann K. Investigation of the effects of the fatty acid profile on fuel properties using a multi-criteria decision analysis. Energy Conversion and Management. 2015;**98**:340-347

[52] Knothe G, Razon LF. Biodiesel fuels. Progress in Energy and Combustion Science. 2017;**58**:36-59

[53] Yaşar F. Comparision of fuel properties of biodiesel fuels produced from different oils to determine the most suitable feedstock type. Fuel. 2020;**264**:116817

[54] Sinha SK, Gupta A, Bharalee R. Production of biodiesel from freshwater microalgae and evaluation of fuel properties based on fatty acid methyl ester profile. Biofuels. 2016;**7**(1):69-78

[55] Nascimento IA, Marques SSI, Cabanelas ITD, Pereira SA, Druzian JI, de Souza CO, et al. Screening microalgae strains for biodiesel production: Lipid productivity and estimation of fuel quality based on fatty acids profiles as selective criteria. Bioenergy Research. 2013;**6**(1):1-13

[56] Moniruzzaman M, Yaakob Z, Shahinuzzaman M, Khatun R, Aminul Islam AKM. Jatropha biofuel industry: The challenges. Frontiers in Bioenergy and Biofuels. 2017;**1**(12):23-256

[57] Lin CY, Ma L. Influences of water content in feedstock oil on burning characteristics of fatty acid methyl esters. PRO. 2020;**8**(9):1130

[58] Krishnakumar U, Sivasubramanian V, Selvaraju N. Physico-chemical properties of the biodiesel extracted from rubber seed oil using solid metal oxide catalysts. International Journal of Engineering Research and Applications. 2013;**3**(4):2206-2209

[59] Atmanli A. Comparative analyses of diesel–waste oil biodiesel and propanol,

n-butanol or 1-pentanol blends in a diesel engine. Fuel. 2016;**176**:209-215

[60] Ateeq M, LiLi MA, Ahmed A, Gohar GA, Rafiq M, Rauf S, et al. Evaluation corrosion effect of biodiesel produced from neem oil on automotive materials. Journal of Materials Today Sustainability. 2022;**18**:100130. DOI: 10.1016/J.mtsust.2022.100130

[61] Ozcanli M, Gungor C, Aydiu K. Biodiesel fuel specifications: A review. Journal of Energy Sources, Part A. 2013; **35**:635-647. DOI: 1080/15567036

[62] Deyab MA. Corrosion inhibition of aluminum in biodiesel by ethanol extracts of Rosemary leaves. Journal of the Taiwan Institute of Chemical Engineers. 2016;**58**:536-541

[63] Available from: https://www.researchgate.net. [Accessed: June 27, 2022]

[64] Hoang AT, Tabatabaei M, Aghbashlo M. A review of the effect of biodiesel on the corrosion behavior of metals/alloys in diesel engines. Energy Sources, Part A: Recovery, Utilization, and Environmental Effects. 2020;**42**(23): 2923-2943

[65] Pugazhendhi A, Arvindnarayan S, Shobana S, Dharmaraja J, Vadivel M, Atabani AE, et al. Biodiesel from Scenedesmus species: Engine performance, emission characteristics, corrosion inhibition and bioanalysis. Fuel. 2020;**276**:118074

[66] Deyab MA, Keera ST. On corrosion and corrosion inhibition of carbon steel in stored biodiesel: Electrochemical (AC and DC) studies. Journal of the Taiwan Institute of Chemical Engineers. 2016; **68**:187-191

[67] Subedi BN, Amgain K, Joshi S, Bhattarai J. Green approach to corrosion inhibition effect of Vitex negundo leaf extract on aluminum and copper metals in biodiesel and its blend. International Journal of Corrosion and Scale Inhibition. 2019;**8**(3):744-759

[68] Suresh M, Jawahar CP, Richard A. A review on biodiesel production, combustion, performance, and emission characteristics of non-edible oils in variable compression ratio diesel engine using biodiesel and its blends. Renewable and Sustainable Energy Reviews. 2018; **92**:38-49

[69] Hasan MM, Rahman MM. Performance and emission characteristics of biodiesel–diesel blend and environmental and economic impacts of biodiesel production: A review. Renewable and Sustainable Energy Reviews. 2017;**74**:938-948

[70] Jakeria MR, Fazal MA, Haseeb ASMA. Effect of corrosion inhibitors on corrosiveness of palm biodiesel. Corrosion Engineering, Science and Technology. 2015;**50**(1):56-62

[71] Karmakar R, Rajor A, Kundu K, Kumar N. A comparative study of the fuel characteristics between algae biodiesel and petrol-diesel. In: Bioresource Utilization and Bioprocess. 1st ed. Singapore: Springer; 2020. pp. 49-55. DOI: 10.1007/978-981-15-1607-8_5

[72] Mishra S, Anand K, Mehta P. Predicting the cetane number of biodiesel fuels from their fatty acid methyl ester composition. Energy Fuels. 2016;**30**(12):10425-10434. DOI: 10.1021/acs.energyfuels.6b01343

[73] Baena LM, Calderón JA. Effects of palm biodiesel and blends of biodiesel with organic acids on metals. Heliyon. 2020;**6**(5):e03735

[74] Riayatsyah TMI, Sebayang AH, Silitonga AS, Padli Y, Fattah IMR, Kusumo F, et al. Current progress of *Jatropha Curcas* commoditization as biodiesel feedstock: a comprehensive review. Frontiers in Energy Research. 2022. DOI: 10.3389/fenrg.2021.815416

[75] Karmakar R, Kundu K, Rajor A. Fuel properties and emission characteristics of biodiesel produced from unused algae grown in India. Petroleum Science. 2017; **15**:385-395

[76] Akhihiero ET, Aluyor EO, Audu TOK. Effect of variation of temperature on the transesterification of Jatropha seed oil using a homogeneous catalyst. Journal of Advanced Material Research. 2013;**824**:473-476. DOI: 10.4028/www.scientific.net/AMR824.473

[77] Akhihiero ET, Oghenejoboh KM, Umukoro PO. Effect of process variables on transesterification reaction of jatropha curcas seed oil for the production of biodiesel. International journal of emerging technology and advanced engineering, ISO 9001:2008 Certified Journal. 2013;**3**(6):388-393

[78] Akhihiero ET, Audu TOK, Kuale PA. Performance characteristics of a locally fabricated hydraulic press for the extraction of oil from oil bearing seeds. Journal of the Nigerian Society of Engineers. 2013;**47**(2):25-28

[79] Sharmbiu VB, Bhattacharya TK, Nayak IK, Das S. Studies on characterization of raw jatropha oil and its biodiesel with relevance of diesel. International Journal of Engineering Technology and Advanced Engineering. 2013;**3**(4):48-54

[80] Sepidar S, Zurina ZA, Robiah Y, Azhari M. Extraction of oil from jatropha seeds, optimization, and kinetics. American Journal of Applied Sciences. 2009;**6**(7):1390-1395

[81] Akhihiero ET. A research note on oil extraction from Jatropha Curcas Seeds. International Review of Chemical Engineering. 2013;**5**(4):301-304

[82] Al-Zuhair S, Ling FW, Jun LS. Proposed kinetic mechanism of the production of biodiesel from palm oil using lipase. Process Biochemistry. 2007; **42**(6):951-960

[83] Murugesan A, Umarani C, Subramanian R, Nedunchezhian N. Bio-diesel as an alternative fuel for diesel engines A review. Renewable and Sustainable Energy Reviews. 2009;**13**(3): 653-662

[84] Akhihiero ET, Oshoma V, Ebhodaghe SO. Kinetics of the fermentation of cassava starch. University of Benin Journal of Science and Technology. 2021;**6**(1)

[85] Meira M, Santana PMB, Araújo AS, Silva CL, Leal Filho JR, Ferreira HT. Oxidative degradation and corrosiveness of biodiesel. Corrosion Reviews. 2014;**32** (3–4):143-161

[86] Ambat I, Srivastava V, Sillanpää M. Recent advancement in biodiesel production methodologies using various feedstock: A review. Renewable and Sustainable Energy Reviews. 2018;**90**: 356-369

[87] Yasar F. Evaluation and advantages of algae as an energy source. Journal of the Turkish Chemical Society. 2018;**5**(3): 1309-1318

[88] Yatish KV, Lalithamba HS, Suresh R, Omkaresh BR. Synthesis of biodiesel from Garcinia gummi-gutta, Terminalia belerica and Aegle marmelos seed oil and

investigation of fuel properties. Biofuels. 2018;**9**(1):121-128

[89] Li Y, Tian G, Xu H. Application of biodiesel in automotive diesel engines. In: Biodiesel - Feedstocks, Production and Applications. London, United Kingdom: IntechOpen; 2012. DOI: 10.5772/53222

[90] Available from: https://afdc.energy.gov/vehicles/diesel.html [Accessed: July 14, 2022]

[91] Available from: https://biofuels-news.com/news/commercial-vehicle-firms-join-new-list-of-biodiesel-compliance/ [Accessed: July 14, 2022]

[92] Hajjari M, Tabatabaei M, Aghbashlo M, Ghanavati H. A review on the prospects of sustainable biodiesel production: A global scenario with an emphasis on waste-oil biodiesel utilization. Renewable and Sustainable Energy Reviews. 2017;**72**:445-464

[93] Muñoz M, Moreno F, Monné C, Morea J, Terradillos JJRE. Biodiesel improves lubricity of new low sulphur diesel fuels. Renewable Energy. 2011;**36** (11):2918-2924

[94] Available from: https://www.theguardian.comFeb.2009 [Accessed: May 27, 2022]

[95] Fazal MA, Sazzad BS, Haseeb ASMA, Masjuki HH. Inhibition study of additives towards the corrosion of ferrous metal in palm biodiesel. Energy Conversion and Management. 2016;**122**: 290-297

[96] Daouda D, Douadi T, Ghobrini D, Lahouel N, Hamani H. Investigation of some phenolic-type antioxidants compounds extracted from biodiesel as green natural corrosion inhibitors; DFT and molecular dynamic simulation, comparative study. In AIP Conference Proceedings. Vol. 2190, No. 1. Melville, NY: AIP Publishing LLC; 2019. p. 020098

[97] Gebremariam SN, Marchetti JM. Economics of biodiesel production. Energy Conversion and Management. 2018;**168**:74-84

[98] Christopher LP, Kumar H, Zambare VP. Enzymatic biodiesel: Challenges and opportunities. Applied Energy. 2014;**119**: 497-520

[99] Pourzolfaghar H, Abnisa F, Daud WMAW, Aroua MK. A review of the enzymatic hydroesterification process for biodiesel production. Renewable and Sustainable Energy Reviews. 2016;**61**: 245-257

[100] Kirubakaran M, Selvan VAM. A comprehensive review of low cost biodiesel production from waste chicken fat. Renewable and Sustainable Energy Reviews. 2018;**82**:390-401

[101] Aransiola EF, Ojumu TV, Oyekola OO, Madzimbamuto TF, Ikhu-Omoregbe DIO. A review of current technology for biodiesel production: State of the art. Biomass and Bioenergy. 2014;**61**:276-297

[102] Bhuiya MMK, Rasul MG, Khan MMK, Ashwath N, Azad AK, Hazrat MA. Second generation biodiesel: potential alternative to-edible oil-derived biodiesel. Energy Procedia. 2014;**61**:1969-1972

[103] Talha NS, Sulaiman S. Overview of catalysts in biodiesel production. ARPN Journal of Engineering and Applied Sciences. 2016;**11**(1):439-442

[104] Rezania S, Oryani B, Park J, Hashemi B, Yadav KK, Kwon EE, et al. Review on transesterification of non-edible sources for biodiesel production with a focus on economic aspects, fuel

properties and by-product applications. Energy Conversion and Management. 2019;**201**:112155

[105] Su CH. Recoverable and reusable hydrochloric acid used as a homogeneous catalyst for biodiesel production. Applied Energy. 2013;**104**: 503-509

[106] Abdullah SHYS, Hanapi NHM, Azid A, Umar R, Juahir H, Khatoon H, et al. A review of biomass-derived heterogeneous catalyst for a sustainable biodiesel production. Renewable and Sustainable Energy Reviews. 2017;**70**: 1040-1051

[107] Ramos M, Dias APS, Puna JF, Gomes J, Bordado JC. Biodiesel production processes and sustainable raw materials. Energies. 2019;**12**(23): 4408

[108] Ruhul AM, Kalam MA, Masjuki HH, Fattah IR, Reham SS, Rashed MM. State of the art of biodiesel production processes: A review of the heterogeneous catalyst. RSC Advances. 2015;**5**(122): 101023-101044

[109] de Lima AL, Ronconi CM, Mota CJ. Heterogeneous basic catalysts for biodiesel production. Catalysis Science & Technology. 2016;**6**(9):2877-2891

[110] Beetil K, Sadly SB, Taleb-Hossenkhan N, Bhagooli R, Puchooa D. An investigation of biodiesel production from microalgae found in Mauritian waters. Biofuel Research Journal. 2014;**2**: 58-64

[111] Khan S, Siddique R, Sajjad W, Nabi G, Hayat KM, Duan P, et al. Biodiesel production from algae to overcome the energy crisis. HAYATI Journal of Biosciences. 2017;**24**:163-167. Available from: https://www.journals.elsevier.com/hayati-journal-of-biosciences

[112] Available from: https://www.globalpetrolprices.com/diesel_prices/ [Accessed: July 14, 2022]

Section 5

Renewable Energy Microgrids

Chapter 9

100 MW Wind Turbine Power Plant

Samuel A. Alagbada

Abstract

Wind power production has increased by a hundredfold during the last 20 years and represents roughly 3% of the total global electricity production. In recent years, technological changes in wind turbine configurations have enabled higher capacity factors for wind turbines. The results from the studies showed that wind as a source of energy for Växjö could be explored in order to achieve the goal of energy sufficiency and as well as sustaining the greenest city status in Europe. The simulation showed that 100 MW electricity could be generated from the wind sources with respect to the available data via global wind metrological data, literature, RETScreen Expert software., LCOE and IRR analysis tools. In addition, the Internal rate of return (IRR) of 8.7% which is good enough considering the proposed energy tax, energy security and environmental benefit cost ratio as well as reduced global weighted-average levelized cost of electricity (LCOE) from wind power technology make it more attractive for investor-Växjö municipality.

Keywords: wind power plant, wind energy, electricity storage, RETScreen expert, LCOE

1. Introduction

The global pursuit of access to affordable, reliable, sustainable, and modern energy for all as a Sustainable Development Goal (SDG) number 7 among 193 member countries are synonymous to clean energy, carbon neutrality, efficient energy management and transition from non-renewable to a renewable source [1]. According to the long-term scenarios of the International Energy Agency identified wind and solar renewable power production as a pathway in achieving the set goals in order to avoid the worst impacts of climate change. The International Renewable Energy Agency predict that wind power will lead the way in the energy transition globally and be the prominent source as a measure to reduce carbon emissions [2].

Wind power industry has become the world's fastest growing renewable energy source [3]. More than 70 countries around the globe including Sweden, contribute to the global wind generating capacity of about 1300 TWh with an average growth rate of 25% per year globally, between the years 1994 and 2015. This correspond to roughly 3% of the total global electricity production as well as saving the planet about 1.1 billion tonnes of CO2 globally as of today [4]. In addition, in a future scenario from the International

Energy Agency (2021), wind power is forecasted to provide 18% of the total electricity globally in year 2050 with over 25 billion dollars investment as of today [4].

Wind power is placed to be one of the cornerstones of green recovery and to play an important role in accelerating the global energy transition. In the European Union, the energy generation by renewable sources has been dramatically increased to reach 740 TWh, where the energy production through wind energy is the dominant sector [5].

The Swedish Transmission System Operator (TSO) saddled with responsibility of managing the national power grid network in Sweden, envisaged the need for rapid growth of the wind power energy into the network as a bridge builder as the planned decommissioning of nuclear power plant by the Swedish government timeline of 2030 approaches [2].

Växjö municipality developed and adopted an Energy plan for Fossil -free municipality with an efficient and effective energy system. Kronoberg Energy plan 2025 (Rus -Grona Kronoberg 2025) strategy was developed at the regional level as an integral platform for the adoption of EU, Swedish national energy plan vision see Appendices 1–3 with an ambition of making the region an "Energy-plus County". The statistical data of the Växjö municipality energy balance sheet with respect to demand, supply and consumption to be 2336GWh (Renewable- 60% & Non-renewble-40%) [6, 7].

In addition, the wind power system is dominated by variable-speed among the installed wind turbines, however, technological deployment of power electronic converter, large scale energy storage regulating power, a geographical distribution of wind turbines and transmission capacity to other regions as well as configuration of the wind turbines for high wind power penetration levels are used to improve power quality [8]. The marginal production cost related to wind power are decreasing, partly driven by technology improvements and legislative treaties to achieve an overall reduction in greenhouse gases [9].

The identified factors which may influence the spot prices are electricity consumption, electricity production, electricity flow, electricity capacity, wind power production and other electricity indexes. The varying energy sport price at different region in Sweden (SE3/SE4) are synonymous to the wind production [10].

The efficiency of both land and offshore Wind power technology is synonymous to the height of the turbines tower reaching better wind conditions, the swept area generator size and rotor diameter coupled with electrical subsystems and control units. Size [9]. The literature available suggests that large-scale (WPPs) can have a significant impact on the grid [1–10].

The aim of this paper is to examine the feasibility of wind power technology in Växjö in relation, economic social and environmentally sustainable goals of being a net exporter of energy for both current and future scenario.

2. Theory

The following sections describes the energy in wind power, wind turbine, electricity storage and some economics of investments.

2.1 Wind power

Wind results from differences in air temperature, density, and pressure from uneven solar heating of the Earth's surface. Wind currents act as giant heat exchangers, cooling the tropics and warming the poles. The average annual wind speed, wind patterns near the ground are critical in selecting the height of the hub (center of the rotor) as well as location criteria for the installation of wind turbines. Wind shear is the change in wind speed with height, which is influenced by solar heating, atmospheric mixing, and nature of the terrain but forests and cities element tend to increase wind shear by slowing the speed of air near the surface.

2.2 Wind turbine

Wind turbine transforms kinetic energy of air currents into electrical energy. The energy is mainly extracted with the rotor, which transforms the kinetic energy into mechanical energy, and with the generator, which transforms this mechanical energy into electrical energy. It is renewable, efficient, mature and secure energy that is key to the energy transition and the decarbonisation of the economy.

A typical wind turbine consists of the following subsystems as follow in **Figure 1** [11].

- Rotor (consists of blades and hub)
- Drive-train (shafts, gearbox, couplings, mechanical brake, and electrical generator)
- Nacelle and main-frame (housing, bedplate, and yaw system)
- Tower and foundation
- Electrical system (cables, switchgear, transformers, and power electronic converters if present)

Wind turbines are classified into two in relation in relation to how its spin. There are two kinds of wind turbines. Wind turbines that rotate along its vertical axis is the vertical axis wind turbines (VAWT), while the ones that spins about a horizontal axis is the horizontal axis wind turbines (HAWT).

2.2.1 Wind turbine energy

Wind energy can be described by the kinetic energy of the particles in the air. The energy content, E_{wind} in a mass of air, m_{air}, and the wind speed v^2 is described as

$$E_{wind} = 0.5.m_{air}.v^2 \tag{1}$$

P_{wind}, can be described by the energy in a volume of air that passes an area, A, and density of air (P_{air}). The density varies with the altitude and time, since it depends on the temperature and pressure.

$$P_{wind}, \; = 0.5.P_{air}.A.v^3 \tag{2}$$

Wind power efficiency is defined as

$$\eta = \frac{8}{27}.P_{air}.A.v^3 \tag{3}$$

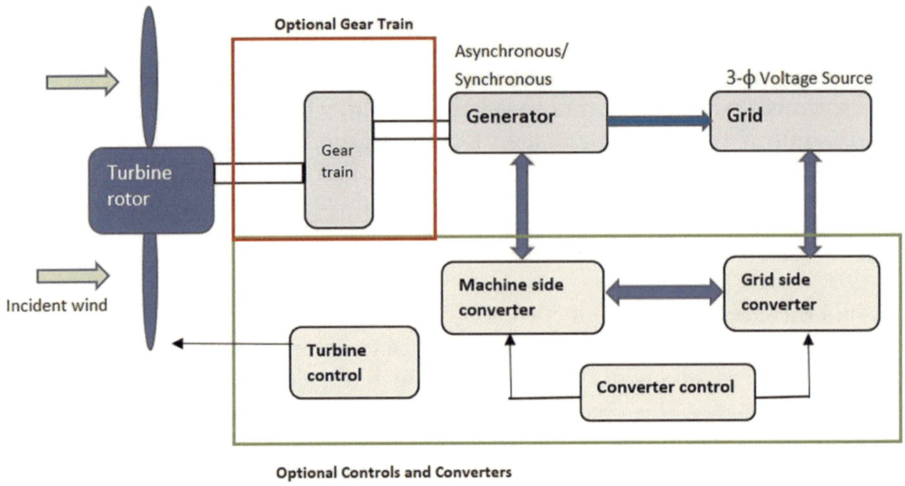

Figure 1.
A schematic diagram of a wind turbine [11].

2.2.2 Wind turbine power

The volume of energy harvested from a turbine is a function of Wind power (P_{wind},) and coefficient of performance (C_p). The coefficient of power is depending on the blade design and its configuration in relation to the blade pitch angle and the tip speed ratio(λ). The Optimal value of C_p is approximately 7 thresholds. Higher value of C_p above the threshold of 7 allows for better efficiency while below the threshold of 7 decrease its efficiency with increased noise, blade erosion, drag losses and increased flow around the wind turbine instead of through it.

A power curve describes the power production of a wind turbine as a function of the wind speed. Power curves could either be provided by the wind turbine manufacturer or be approximated. An approximated cubic wind power function curve, P_{cub}can be modelled for all wind speeds.

$$P_{real}(v) = \begin{cases} 0 \\ 0.5\,.C_p\ max\,(v).P_{air}.A.v^3 \\ p_r \end{cases}$$

$$\begin{array}{l} if\ v_{ci}\ > v\ or\ v > v_{co} \\ if\ v_r\ \geq v\ \ \geq v_{ci} \\ if\ v_{co}\ \ \geq v\ \ \geq v_r \end{array} \qquad (4)$$

where p_r is the rated power, v is the rated wind speed, v_{ci} is the cut-in wind speed and v_{co} is the cut-out wind speed usually around 3 m/s and 25 m/s respectively. A higher capacity factor and hub height are important factor for the electricity production from wind turbines.

Wind resource is a good tool for measuring the historical wind power production, wind speed at the hub height. In modelling for installation of a wind turbine, wind, a wind shear relationship is used for the data correlation at desired altitude for optimal efficiency.

$$\frac{v_1}{v_2} = \left(\frac{h_2 - h_{disp}}{h_1 - h_{disp}}\right)^{\alpha} \tag{5}$$

where v_1 is the measured wind speed at the height h_1, v_2 is the unknown wind speed at the hub height h_2, h_{disp} is the displacement height, which is the height at which the wind speed is projected to be zero for modelling purposes, and α is the wind shear exponent, which is depending on the ground roughness.

2.3 Electricity storage

The wind power has operational risk from sudden changes in the weather that cuts supply and adversely affects grid stability. The intervention by dispatchers is alternative to compensate for the inherent unreliability of wind power. Grid storage transforms the highly variable nature of renewable power to a reliable supply under the direct control of dispatchers.

Grid storage acts as an energy buffer supplying power for current use and absorbing excess renewable power for later use. It creates value by storing wind electricity when it has little or no value and feeding it into the electricity grid when it does have value in a night and day scenario. It also acts as a grid stabiliser enhancing operating efficiency as shown in Appendix 4. The primary means today are pumped storage or gravity batteries, while the future opportunities lie with non-flow and flow batteries, with compressed air and hydrogen.

2.4 Economics

The economical potential of an investment in energy in relation to the present, and future scenarios could be analysed using economic indexes such as cash flows, the discount rate, inflation, levelized cost, risks and uncertainties of the investment. The basis for investment decisions can be made by including the variations in spot price of energy.

$$c_{inv} = c_{inv}.\frac{r.(1+r)^y}{r.(1+r)^y - 1} \tag{6}$$

The levelized cost of electricity from a wind turbine, LCOE, expresses an equally shared production cost per unit of electricity produced by the turbine. The expression for

$$LCOE = C_{varO\&M} + .\frac{C_{inv} + C_{fixO\&M}}{AEP} \tag{7}$$

where AEP is the annual electricity production, $C_{varO\&M}$ is the variable operation and maintenance ($O\&M$) costs, and there are no fuel costs for wind power..

The yearly net profit, $Profit_{net}$ from wind power production is defined in eq. (8). However, without taking into account regulating costs, taxes, other costs or subsidies

$$Profit_{net} = \sum_{k=1}^{8760} \left(\left(P_{spot}(t) - C_{varO\&M} \right) . P_{wind}(t) \right) - C_{inv} - C_{fixO\&M} \tag{8}$$

The profitability of an investment can be measured by the yearly return on investment, ROI,

$$ROI = \frac{Profit_{net}}{C_{inv}} \quad (9)$$

3. Method

The following chapter describes the experimental methods used to collect and analyse the data for the purpose of this paper.

This section introduces system engineering method as shown in Appendix 5 in modelling the a 100 MW capacity wind power plant for vaxjo location with geographical information shown in Appendix 6. The wind power value study is made on current power systems, electricity to be produced from wind power for the present and future in connection with a social-economic perspective.

Wind power production is estimated using wind turbine configurations, wind data from Global wind atlas [12] see Appendices 7 and 8, height of 100 M, a power curve model and other parameter values according to Eq. (1)–(5) were obtained from the Ge turbine manufacturer database [13, 14]. The Investment, Operational & maintenance costs for the wind turbines and other financial variables for the power plant is obtained from RETScreen Expert software. Post simulation analysis for LCOE and ROI according to Eqs. (7)–(9) were carried out.

The model is simulated with the RETScreen Expert software. RETScreen Expert software is a clean energy management system which intelligently enables professionals and decision-makers to rapidly identify and assess the viability of potential energy efficiency, renewable energy and cogeneration projects; and to easily measure and verify the actual and ongoing energy performance of buildings, factories and power plants around the world.

3.1 Data and assumptions

This section contains explanations of the data and the assumptions used in this case studies. The wind turbine model (50 numbers of GE Wind 1.5 s wind turbine, hub height 80 m, rotor diameter per turbine 70.5 m and swept area per turbine 3,903 m^2. The density of air is set constant at 1.225 kg/m^3. for all heights with capacity factor of 25% with 42% efficiency factor. The cut-in and cut-out speeds are set to 3m/s and 22.5m/s, respectively as shown Appendix 9, and the wind speeds are normalised around the average wind speed, with a standard deviation of 1m/s for the regional aggregation. The power curve of the turbine from the GE electric of the turbine are shown in Appendix 7.

The LCOE calculation assumptions: The Initial cost is set to $21,000,000 and O&M cost $7,000,000 used are those stated by the GE vendors. The economic lifetime is set to the technical lifetime of 30 years as shown in Appendix 10 [15, 16] and discount rate for the wind value study is set to 3.0% in line with other publications in this field [16]. Electrical output is assumed to be 50 Hz (European electricity grids).

Wind power plant of 100 MW production target and investment		
Investment cost—Year 0 ($)	21,000,000	
O&M cost ($)	7,000,000	
O&M growth rate (%)	0,0%	
Fuel cost—1st year ($)	0	
Fuel cost growth rate (%)	0,0%	
Project lifespan (year)	30	
Discount rate (%)	3,0%	
Summary		
Electricity export to the grid (MWh)		219,898
Electricity export Revenue ($)		21.989,775
GHG emission reduction tco_2		2.925
Payback period		16
ROI		8.7
LCOE ($/MWh)		90

Table 1.
Wind power plant production target and investment.

4. Results

In this section, the outcomes of the experiment result data were gathered for analysis.

This study assesses how to optimise the revenue from wind power, in order to make profitable investment decisions through the choice of wind turbine technology. **Table 1** showed Wind power plant electricity production target, investment ROI and LCOE. The project cash flow curve, Payback period are shown in **Figure 2**.

5. Discussion

In this section, the outcomes of the experiment result data were analysis and interpreted.

5.1 Choice of GE turbine

GE Renewable Energy is one of the world's leading wind turbine suppliers, with over 49,000 units installed and generating wind electricity across the globe in providing wind energy solution that best address your challenges and priorities irrespective of the power plant location of either onshore or offshore. The high wind energy reliability as well as wind power efficiency with the use of a broad family of smart, modular turbines that are uniquely suited for a variety of wind environments has been their strength in the global player in the energy market [15]. Therefore, the choice of GE turbine for this study is appropriate in relation to Appendix 11.

5.2 CO_2 emissions

The Swedish energy security scenario from 1960s up till 2018 has being implemented via an environmentally sustainable policies such as Carbon tax,

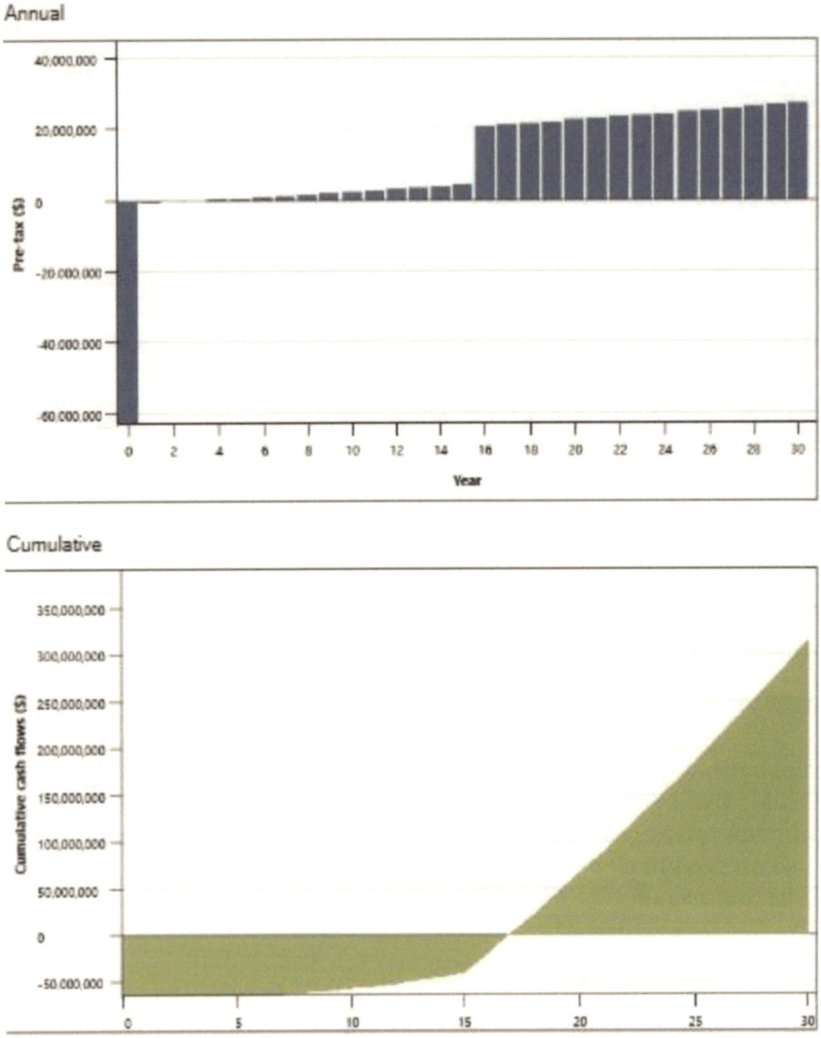

Figure 2.
Project cash flow.

transition from non-renewable ton renewable source of energy, GHG reduction, and climate neutrality. Sweden has the highest carbon tax globally and it is charged or levied per tonne of CO2 emissions on the different fuels [17, 18]. With the proposed 100 MW wind power plant, 2.925 tons of CO_2 emission would be avoided.

5.3 The social cultural norms and political will

The government and societal cultural new norms of being sustainable across board from national level to municipalities are collaborating in the pursuit of carbon neutrality goals by adopting and implementing measures, policies, legislation, and issuances of licences to potential investor in translation from non-renewable to renewable sources coupled with energy efficient system for the present and future scenario [6].

Wind power plant in Växjö would align with municipality energy plan surplus goals for export, and remain the greenest city in Europe.

5.4 Technology developments

The KTH Swedish royal institute of technology study concluded that vaxjo had a technical potential estimated of almost 9000 GWh electricity could be generated from wind power [6]. Globally, a great deal of technological progress has been made in lowering the levelized cost of electricity generated from solar and wind [19, 20] . The technical solution are evolving for the development of a super battery that can store wind electricity generated when it's not needed and act as an energy bank for dispatching when needed, also, smoothing its output as a solution to the reliability setback of wind power [21, 22]. Therefore, the SMART acronym of sustainability, reliability and affordability in the energy field could be resolved within the short- and long-term niche of technology advancement.

5.5 Economics

The Wind power plant of 100 MW with all related economical value of investment and plant production target as shown in **Table 1**. Levelized Cost of Energy (LCOE) being the minimal electricity production costs and is calculated for assessing different sources of power generation [23]. World Energy Transitions Outlook 2021 envisaged that very low-cost renewables for a speedy delivery decarbonised electricity system [24]. The global installation cost of onshore wind turbine and LCOE has being reduced by 31% and 56% respectively [19]. EIA 2020 estimates of the global Levelized cost of electricity (LCOE) and levelized cost of storage (LCOS) for onshore wind power plan has been 56$/MWh with 10% discount rate [20]. Also, the LCOE values would be same irrespective of the size [25]. therefore, 90$/MWh obtained from feasibility studies could be related to onshore wind power plant database from both IRENA & EIA. This study result seems are within permissible range.

6. Conclusions

The needs to be installing wind power faster over the next decade in order to stay on a net zero pathway and avoid the worst impacts of climate change is necessary.

Wind being a clean energy technology with the most decarbonization potential per MW, coupled with economies of scale in relation to the pursuit of the vision 2050 and net exporter of energy has made investment in wind power to be more attractive.

The results shows that wind power can play a major role in future power systems with zero carbon emissions.

For an optimal wind power penetration, the annual wind speed, hub height ground roughness and electricity spot price are akin to investment decision in relation to LCOE.

For the proposed 100 MW wind power plant LCOE stood at 90$/MWh with IRR of 8.7% are feasible for an onshore wind power plant in Växjö on a social economical font so that the net exporter of electricity and energy security goal of the municipality could be accomplished and sustained.

However, a sensitivity analysis is necessary to determine how potential changes against electricity spot market prices under certain circumstances in the future scenario.

Acknowledgements

My profound gratitude goes to all who has being part of this journey from inception.

A. Appendix 1: Swedish energy policy

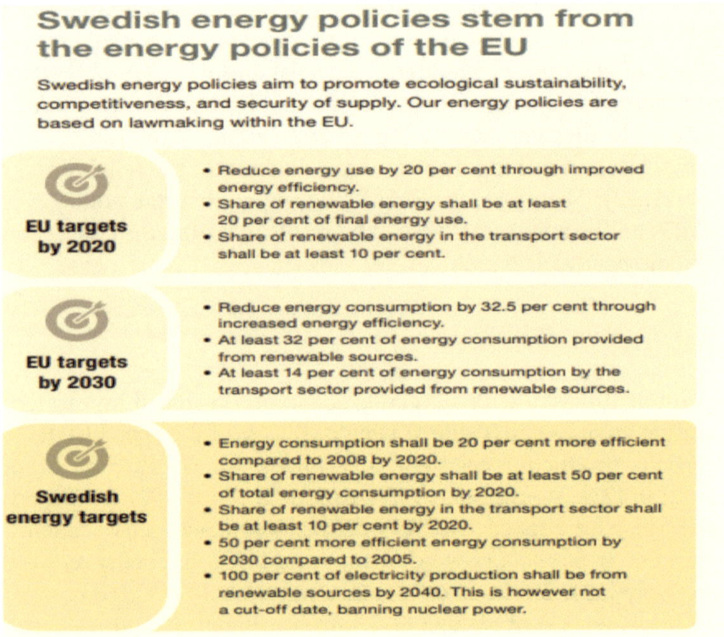

http://www.swedishenergyagency.se/.

B. Appendix 2: Swedish energy system 2019

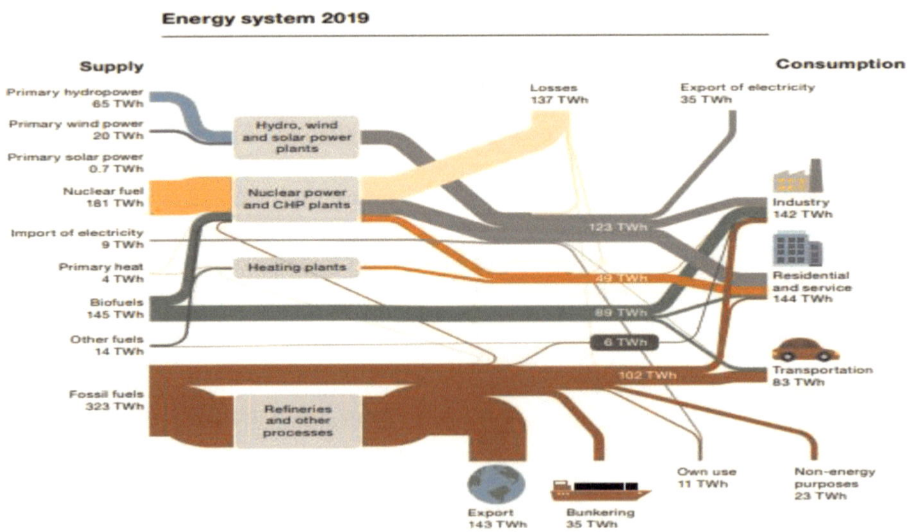

http://www.swedishenergyagency.se/.

C. Appendix 3: Swedish Total energy supplied 1970–2019

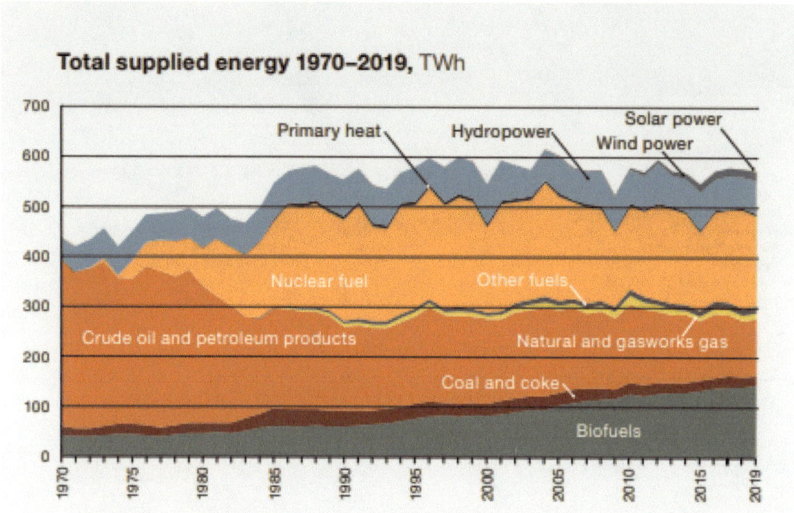

http://www.swedishenergyagency.se/.

D. Appendix 4: Performance of storage technologies

See **Table A1**.

Storage type	Power in mW	Discharge Time	Efficiency %	Lifetime Years	Storage cost $/mW-hour
Pumped storage	250–1000	10 hrs	70–80	<30	$50–150
CAES	100–300	3–10 hrs	45–60	30	About $150
Flywheels	<10	<15 min	>85	20	NA
Supercapacitors	10	<30 sec	About 90	50 k cycles	NA
Vanadium RB	<10	2–8 hrs	75	5–15	$200–300
Lithium ion	5	<4 hrs	About 90	8–15	$250–500
Lead battery	3–20	<4 hrs	75	4–8	NA
Sodium sulfur	30–35	4 hrs	80	15	$50–150

Table A1.
Performance of storage technologies.

E. Appendix 5: System engineering model

See **Figure A1**.

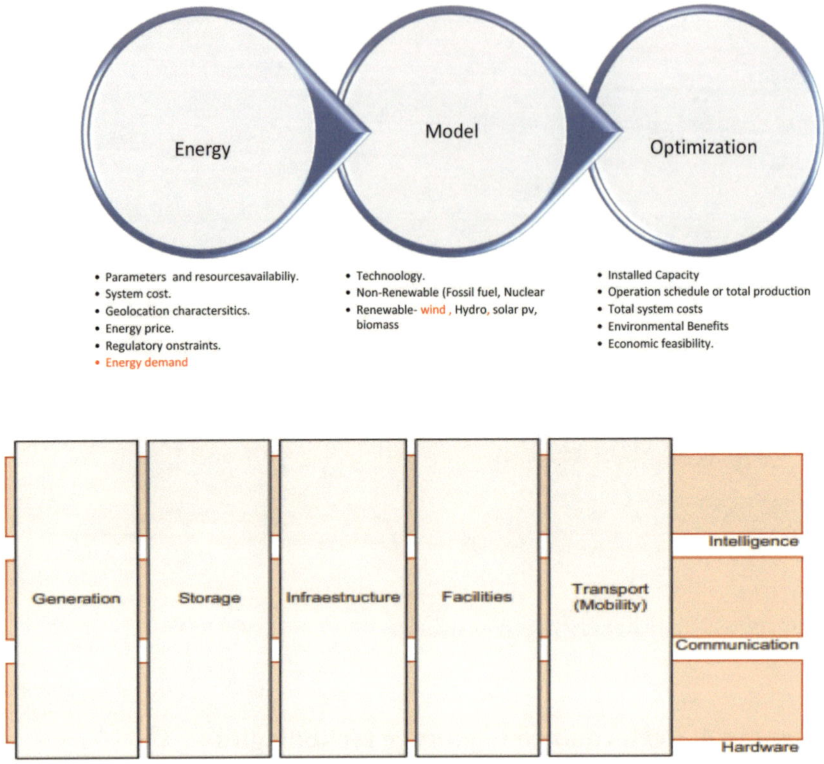

Figure A1.
Classification of energy intervention areas in the smart city.

F. Appendix 6: Locational data for Vaxjo Res screen Expert

See **Figure A2**.

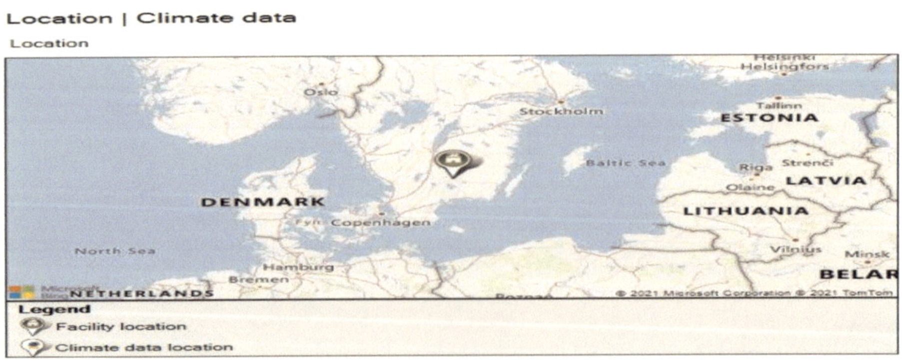

	Unit	Climate data location	Facility location
Name		Sweden - Vaxjo/Kronoberg	Sweden
Latitude	°N	56.9	56.9
Longitude	°E	14.7	14.7
Climate zone		6A - Cold - Humid	6A - Cold - Humid
Elevation	m	186	176

Figure A2.
Locational data for Vaxjo Res screen expert.

G. Appendix 7: Power Curve of The wind turbine GE Wind 15 s

See **Figure A3**.

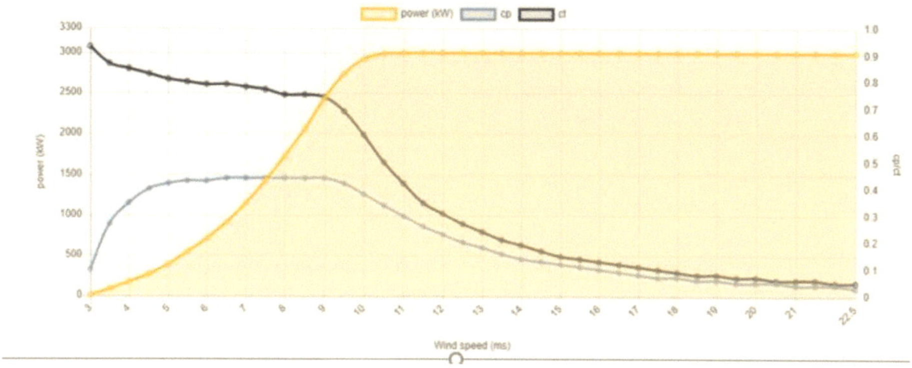

Figure A3.
Power Curve of The wind turbine GE Wind 15s.

H. Appendix 8. Växjo mean power density curve (wind global atlas)

See **Figure A4**.

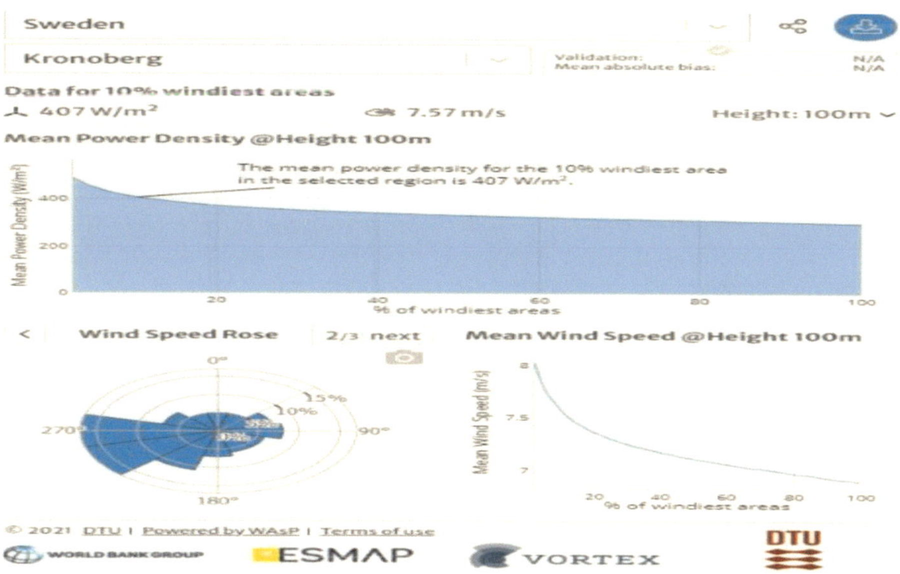

Figure A4.
Växjo mean power density curve (wind global atlas).

I. Appendix 9. Wind speed classification for wind turbines

See **Table A2**.

Wind speed (mph)	Power output	Comments
0–5	None	Wind speed too low to produce power
5–20	Increases by the cube of wind speed up to 100% at 20 mph	Power output=(wind speed/20)^3×100%
20–40	100% rated output	Turbine operating at maximum performance
40–60	30% rated output	Power is reduced to protect the integrity of the wind turbine from high winds
Over 60	0% rated output	Objective is to preserve the wind turbine structure from damage, not generate power

Table A2.
Project table.

J. Appendix 10. Wind and solar Capacity life tine

See **Table A3**.

Units	Lifetime years	Electrical capacity	Efficiency correction factor
Wind in 2030 (All scenarios)	25	172	0.31
Wind in 2050 (EU & Additional)	30	210	0.31
Wind in 2050 (Electrification)	30	259	0.31
Roof top PV (All scenarios)	40	70	0.13

Table A3.
Wind and solar capacity life time [15].

K. Appendix 11. Global wind turbine manufacturer companies

See **Table A4**.

Company	Percent Market Share	Description
GE Energy (US)	15.5	16,000 1.5 mW installed plus 2.5 mW plus 4.1 mW built for offshore
Vestas (Denmark)	14.0	43,000 turbines in 66 nations including 1.8 and 2.1 mW for onshore, 3 mW for low wind conditions plus 3 mW for offshore via MHI Vestas joint venture
Siemens (Germany)	9.5	Market leader offshore turbines 2.3 mW, 3.6 mW, and 6 mW with 150 meter diameter rotor

Company	Percent Market Share	Description
Enercon (Germany)	8.2	20,000 turbines worldwide including 2 mW, 2.5 mW (light winds), 3 mW, and 7.5 mW
Suzlon Group (India)	7.4	21.5 gW of installed capacity including 2.1 mW for normal and light winds. German subsidiary is REpower 1.8–6.15 mW
Gamesa (Spain)	6.1	18 gW of installed capacity including 2 mW, 4.5 mW, and 5 mW (offshore)
Goldwind (China)	6.0	World's leading manufacturer of permanent magnet direct drive turbines including 1.5 mW and 2.5 mW; other leading Chinese manufacturer is Sinovel

Table A4.
Global wind turbine manufacturer companies [12].

Author details

Samuel A. Alagbada
Linnaeus University, Växjö, Sweden

*Address all correspondence to: aa224nr@student.lnu.se

References

[1] Brown TW, Bischof-Niemz T, Blok K, Breyer C, Lund H, Mathiesen BV. Response to 'burden of proof: A comprehensive review of the feasibility of 100% renewable-electricity systems. Renewable and Sustainable Energy Reviews. 2018;**92**:834-847. DOI: 10.1016/j.rser.2018.04.113

[2] ESIG. Toward 100% Renewable Energy Pathways: Key Research Needs. 2019. Available from: www.esig.energy [Accessed: December 2019]

[3] Cotrell J, Stehly T, Johnson J, Roberts J, Parker Z, Scott G, et al. Analysis of Transportation and Logistics Challenges Affecting the Deployment of Larger Wind Turbines: U.S. Department of Energy Technical Report NREL/TP-5000-61063. NREL; 2014. Available from: www.nrel.gov/publications

[4] Global Wind Report. 2021. Available from: https://gwec.net/global-wind-report-2021/

[5] EU. Methodology for the European resource adequacy assessment in accordance with Article 23 of Regulation (EU) 2019/943 of the European Parliament and of the Council of 5 June 2019 on the internal market for electricity. 2019. Available from: https://www.acer.europa.eu/Official_documents/Acts_of_the_Agency/Individual%20decisions%20Annexes/ACER%20Decision%20No%20242020_Annexes/ACER%20Decision%2024-2020%20on%20ERAA%20-%20Annex%20I.pdf

[6] Energy plan for Växjö Municipality. 2020. Available from: https://vaxjo.se/download/18.7b0b9141162a514f1b113069/1523263213740/Energiplan%202016_eng_webb.pdf

[7] City of Växjö. Welcome to Växjö-The Greenest City in Europe. 2007. Växjö Energi. Available from: https://www.veab.se/om-oss/anlaggningar/ [Accessed: April 30, 2020]

[8] Lysova E. Electricity Spot Price Forecasting in Two Swedish Regions: Analysis of Factors Which Cause Price Differences between SE3 (Stockholm) and SE4 (Malmö) Price Region. Sweden: Linnaeus University, Faculty of Technology, Department of Mathematics; 2015

[9] Wiser R et al. Long-term implications of sustained wind power growth in the United States: Potential benefits and secondary impacts. Applied Energy. 2016;**179**:146-158

[10] Göransson L. The Impact of Wind Power Variability on the Least-Cost Dispatch of Units in the Electricity Generation System [Thesis]. Sweden: Chalmers University of Technology; 2014

[11] Letcher T. Wind Energy Engineering: A Handbook for Onshore and Offshore Wind Turbines. Academic Press; 2017. pp. 155-158. ISBN-9780128094297

[12] Global Wind Atlas. Available from: https://globalwindatlas.info/

[13] Dobschinski J, Kanefendt T. Freileitungsmonitoring auf Basis meteorologischer Informationen. In: Conference proceedings "Zukünftige Netze". Berlin, Germany; 2020

[14] Dobschinski J, Siefert M, et al. Development of innovative weather and power forecast models for the grid integration of weather dependent energy sources. In: Proceedings of the WindAc

Conference, Cape Town, South Africa; Nov 2016

[15] Nersesian RL. Energy Economics Markets, History and Policy. New York: Routledge Taylor & Francis Group; 2016

[16] Razzaq N, Muhammad F, Karim R, Tariq M, Muhammad K. The nexus between energy, environment and growth: Evidence from Latin-American countries. International Journal of Energy Economics and Policy. 2021;**11**(1):82

[17] Hammar H, Åkerfeldt S. CO2 Taxation in Sweden—20 Years of Experience and Looking Ahead. Stockholm, Sweden: The Swedish Ministry of Finance; 2011

[18] Samuel J, Anders Y, et al. Looking Back on 30 Years of Carbon Taxes in Sweden. FISCAL FACT No. 727 Sept. 2020. Available from: https://taxfoundation.org/sweden-carbon-tax-revenue-greenhouse-gas-emissions/

[19] IREN. Renewable-Power-Costs. 2020. Available from: https://www.irena.org/publications/2021/Jun/Renewable-Power-Costs-in-2020

[20] IEA. Projected Costs of Generating Electricity. 2020. Available from: https://www.iea.org/reports/projected-costs-of-generating-electricity-2020

[21] Holttinen H, Kiviluoma J, et al. Design and Operation of Energy Systems with Large Amounts of Variable Generation: Final Summary Report, IEA Wind TCP Task 25. Finland. VTT Technology No. 396: VTT Technical Research Centre of Finland; 2021. DOI: 10.32040/2242-122X.2021.T396

[22] Calvert S, Thresher R, Hock S, Laxson A, Smith BUS. Department of Energy Wind Energy Research Program for low wind speed Technology of the Future—Discussion. Journal of Solar Energy Engineering Transactions ASME. 2002;**124**:455-463

[23] Rhodes JD et al. A geographically resolved method to estimate levelized power plant costs with environmental externalities. Energy Policy. 2017;**102**: 491-499

[24] EIA. Annual Energy Outlook. 2021. Available from: https://www.eia.gov/outlooks/aeo/pdf/electricity_generation.pdf

[25] Nguyen T. Class Lecture. Växjö, Sweden: Linnaeus University; 2021

Chapter 10

Tracking Trends for Offshore Wind Energy Industries and Infrastructures in the South Korea: Focused on the Jeonnam Shinan 8.2GW and Ulsan 6GW Offshore Wind Farm Projects

Geon Hwa Ryu, Ji Ye Park, Ah Reum Lee, Young Gon Kim and Chae Joo Moon

Abstract

With the international trend of promoting eco-friendly renewable energy for carbon neutrality and the Paris Agreement, South Korea is focusing its national energy mix on renewable energy. Especially, offshore wind energy will be expanded a total power capacity of 12 GW by 2030, which is expected to become South Korea's most important energy source and industrial dynamic force in the future. With the support of the Korean government, many domestic and foreign developers are taking the lead in developing fixed/floating offshore wind energy projects and O&M technology. Through this chapter, we would like to introduce the current status of offshore wind energy in the South Korea, support policies, infrastructure, and issues up to the first quarter of 2022.

Keywords: offshore wind energy, energy business license, offshore wind farm support port, floating offshore wind farm, collector bus

1. Introduction

Offshore wind energy is a representative renewable energy source that uses the kinetic energy of the offshore wind to produce electricity [1, 2]. It is a climate change response technology. Offshore wind energy is a representative climate change response technology that is abundant, continuously renewable, distributed over a wide area, and does not emit greenhouse gases during operation [2]. Therefore, offshore wind energy is the center of global energy trends, including South Korea, and is a major energy source for replacing fossil energy as the core of the energy transition policy [3–5].

South Korea had an early interest in offshore wind power based on the world's best shipbuilding and offshore plant technology, but it was not easily pursued due to the lack of professional manpower, infrastructure, and opposition from environmental groups and fishermen [6, 7]. In recent years, the company is focusing on creating an ecosystem for the offshore wind energy market by shifting from the existing top-down unilateral government-led method to a bottom-up-type growth model based on public-private partnerships. In addition, in accordance with the analysis that large-scale project-oriented supply expansion is necessary to achieve the 12GW target of offshore wind power by 2030, ultra-scale base projects are being carried out in Jeonbuk (8.7GW), Shinan (8.2GW), Ulsan (6GW), and Chungnam (4GW).

In particular, the Shinan 8.2GW project, the world's largest offshore wind farm in a single region, was promoted as a core energy policy by the previous government, and the new government that was just launched in April 2022 announced that it would continue to push forward as long as there are no major problems. Ulsan, belongs to the eastern sea of South Korea, and the eastern sea is very deep, so floating offshore wind projects are being intensively promoted [8, 9]. In general, when the water depth exceeds 50 m, the economical efficiency decreases because the size of the substructure supporting the turbine must be increased [10]. Therefore, in the deep water area, floating offshore wind projects that floats the wind turbine on the water are promoted. A comparison of fixed and floating offshore wind power is shown in **Table 1**. In Ulsan, global companies such as Equinor, Shell-CoensHexicon, CIP/COP, GIG, and RWE are developing or planning to participate in the floating offshore wind project.

As mentioned above, South Korea is being evaluated as a land of opportunity to establish a very good partnership with domestic and foreign offshore wind project developers and wind farm operators. In this chapter, we aim to help domestic and foreign developers and research institutes to become more interested in and obtain information about South Korea's offshore wind energy. The chapter structure is as follows.

- Offshore Wind Energy Policies and Implementation Status in the South Korea
- Post EBL Offshore Wind Farm Projects in the South Korea
- Offshore Wind Farm Support Ports and Hinterlands

Type		Positive	Negative
Onshore Wind Power		• Short Construction Period • Low Price for Installation • Easy Maintenance	• Noise, Transport, Environment, Complaints
Offshore Wind Power	Bottom-Fixed type	• Easy for Installation • Low Price for Maintenance • Can be Large-scale	• Damage to Fishing and Coast Ecosystem • High Price for Installation
	Floating type	• Less fishing because it is far from coast • Can be Large-scale	• High Price for Grid and Maintenance, • Difficult to Install at exceed 100 m water depth

Table 1.
Comparison of advantages/disadvantages of each wind farm method.

In the first subsection, it contains contents to introduce the government's policies and current status related to offshore wind energy in South Korea. In the second section, we introduce the status of acquisition of energy business licenses (EBL) for offshore wind power projects in South Korea as of the first quarter of 2022. Next, we introduce the offshore wind power clusters and hinterlands selected by the West Sea/South Sea/East Sea region of Korea.

2. Offshore wind energy policies and implementation status in the South Korea

Although Korea has a high proportion of carbon-emitting industries and a high dependence on trade with overseas, the pace of change in energy transition is rather slow [11]. In the 'Renewable Energy 3020 Implementation Plan' (2017.12.), Korean government announced the goal of achieving 20% of the renewable energy generation by 2030 [1]. Among renewable energies, it is announced a power generation plan (2020.07.) to complete 12GW offshore wind power capacity by 2030 [1]. Among local governments, Ulsan City, in particular, has set a goal of creating a 6GW large-scale floating offshore wind farm by 2030, and major overseas developers are participating in the projects.

In the detailed section below, we will introduce the roadmap for Korea's offshore wind power and introduce the most representative offshore wind power projects, the Shinan 8.2GW offshore wind power project and Ulsan 6GW floating offshore wind power project.

2.1 Offshore wind power roadmap in the South Korea

Korea's offshore wind power roadmap is based on the Renewable Energy 3020 Implementation Plan mentioned above. The Renewable Energy 3020 Plan means that Korea plans to build a total of 63.8 GW of renewable energy facilities to increase the proportion of renewable energy generation to 20% by 2030. The government promoted large-scale national projects to increase new and renewable energy power generation facilities, and at the same time introduced a local government-led planned location system for the participation of many people [12, 13]. This is a system to encourage resident participation in renewable energy projects and to share profits. In addition, the Korean government is supporting companies in various ways to improve the supply conditions for renewable energy and to foster a new energy industry. **Figure 1** shows the roadmap for offshore wind power in Korea.

The reason that Korea's offshore wind power development has been limited and slow is because there were limitations in four stages. This means that special difficulties existed in each of the stages of site development, resident acceptability, licensing, and project implementation [14–16].

First of all, in the site development stage, problems sometimes occurred in the process of site selection, meteorological mast or LiDAR installation, and acquiring energy business license. Some developers neglected their responsibilities in the process of conducting a wind resource assessment, investing the fishery situation, and prior consultation with residents, and poor preparation led to a slump in the business. In addition, some developers sold energy business licenses to other developers after occupying a project site, which had an impact on the rapid spread of negative

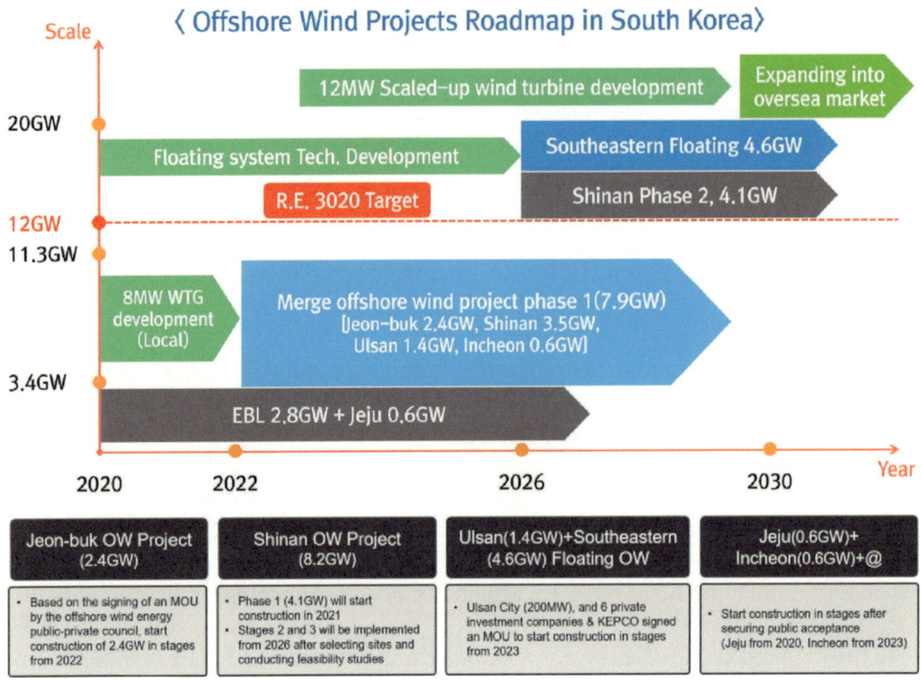

Figure 1.
Offshore wind projects roadmap in South Korea.

perception among residents in the area [3]. These negative perceptions also hindered the good project developers around them.

The second is the problem in the stage of accepting residents. In South Korea, there is a structure in which project developers must independently achieve resident acceptability, but in most cases, there is a large difference in position between project developers and residents. This is because, while residents think that developers neglect their fishing damages, developers think that residents make excessive demands other than legal compensation for damage. In particular, the institutional support mechanism of the state to secure resident acceptance is insufficient, and the government has not intervened in the offshore wind power project because it is a privately-led project, resulting in the sluggishness of the project being neglected for a long time.

The third is the licensing issue. This is because, in the case of civil complaints such as resident acceptability or environmental issues, it is impossible to obtain licenses and permits. Local governments, which are the main licensing authority, cannot actively stand on the side of project developers and ignore residents.

Lastly, there is a problem in the project promotion stage. In the meantime, the formation of the initial domestic market was delayed because it took too long and cost to secure residents' acceptance for offshore wind farm. As a result, a number of machinery/shipbuilding companies withdrew from the offshore wind power projects, which resulted in project developers ignoring domestic wind power companies whose technological development was sluggish.

Accordingly, in July 2020, the Korean government announced an offshore wind power roadmap that could coexist with residents and coexist with the fishery industry through consultation with various ministries. It introduced a total of three initiatives and announced that it would create a 12GW offshore wind farm and create about 87,000 jobs. The three initiatives are as follows.

1. Government-led project discovery and simplification of licensing.

2. Securing resident acceptance and strengthening the environment.

3. Reinforcement of industrial competitiveness in connection with large-scale projects.

In the first implementation plan, the government produces a constraint map for offshore wind farm to produce a digital map for public web service. Through this, areas with good business potential and little damage to fishing are selected and developed as offshore wind power consideration zones under the government. In this process, the government improves the offshore wind power licensing process to support rapid project development, and establishes an integrated licensing organization such as One Stop Shop in Denmark [17].

The second is to introduce a system that can secure resident acceptability and minimize environmental damage. Provide reasonable compensation to the residents of the area around the offshore wind farm, and promote a project that guarantees mid- to long-term income through the participation of the residents. The government will provide guidelines to ensure fishing activities in the vicinity of offshore wind farms and to share business profits. In the process of constructing an offshore wind farm, it must be applied construction methods and products that can minimize environmental damage, and conduct environmental impact analysis by obligating marine environment monitoring. In addition, regulations will be prepared for project developers to deposit a guarantee to restore the marine environment to its original state after the completion of the project.

Lastly, large-scale projects will be prioritized to revitalize the offshore wind energy market, and KEPCO will build a joint connection (Collector bus) facility [18]. The government supports the strengthening of domestic industrial competitiveness, and prepares a roadmap not only for R&D but also for expanding infrastructure such as hinterland ports and wind turbine components test beds [18]. The above three items are made up of several detailed plans and are being pursued sequentially until 2025.

2.2 Shinan 8.2GW offshore wind farm project

Jeollanam-do (Jeonnam) is a metropolitan autonomous region in the southwest of Korea. Jeollanam-do has a goal of building an 8.2GW offshore wind farm in the Shinan region as a core policy of the "Jeonnam Blue Economy", and the national and local governments are making generous investments to create an offshore wind power cluster in the long term. 48.5 trillion won ($38 billion) will be invested over 10 years from 2020 to 2030 [2]. Projects such as the creation of offshore wind farm, the construction of a wind turbine production and assembly complex, the support hinterland, and the construction of transmission lines are carried out in stages. In particular, 2.3 trillion won ($1.8 billion) (investment by KEPCO and power generation companies) will be invested in constructing the transmission line for infrastructure construction, and 218 billion won ($170 million) (Ministry of Oceans and Fisheries) will be invested in the development of the Mokpo New Port [4, 7].

The Shinan offshore wind farm project is a key project in the government's energy transition policy, such as '2050 carbon neutrality' and 'Korean version of the Green New Deal'. The southwest region of Jeollanam-do has excellent potential offshore wind energy and is evaluated as an optimal location for a large-scale offshore wind farm due to its high linkage with shipbuilding and steel, the main industries. In addition, it is revitalizing the local economy by creating large-scale jobs and creating an offshore wind power industry ecosystem with a profit-sharing type win-win job model in which labor, management and the civil government participate together [19]. Korea Electric Power Corporation (KEPCO), Jeonnam Development Corporation, SK E&S, and Hanwha E&C, and many power generation companies are investing in the offshore wind power clusters such as Shinan-Aphae Industrial Complex, Daeyang General Industrial Complex, Mokpo New Port, and Yeongam Daebul Industrial Complex. This is expected to create the wind power industry market and create 117,000 jobs.

The Sinan 8.2GW offshore wind power project is planned to be carried out sequentially in three phases by 2030. If completed, it can become the world's largest single-region offshore wind farm, and the first phase is to build a 4.1GW wind farm from 2020 to 2025. 22 trillion won ($17 billion) is invested in the project, and 1500 new jobs are planned. KEPCO and its subsidiaries will create 1.8 GW, and private power companies will create 2.3 GW. The second phase includes a plan to build a wind farm with a capacity of 2.1 GW from 2022 to 2027, and the 2GW third phase is planning from 2024 to 2030 (**Figure 2**). The support port behind the Shinan 8.2GW project is the Mokpo New Port, which previously handled automobiles, containers, steel, and cement. However, some port areas are being remodeled to be used as piers exclusively for offshore wind power.

Based on the Mokpo New Port and the Daeyang industrial estate, an offshore wind power logistics platform and industrial platform are established to support offshore wind farms to be built in the far and near southwestern seas of Korea and to build an offshore wind farm support cluster. Local governments such as Shinan, Mokpo, Naju, Gwangju, and Yeonggwang form regional clusters to strengthen the competitiveness of the wind power industry at home and abroad and promote local job revitalization. More details will be introduced in Chapter 4.

2.3 Ulsan 6GW floating offshore wind farm project

The government has set a goal of procuring 20% of total power generation from renewable energy, including 12GW of offshore wind power by 2030. Considering that Korea's offshore wind farms currently have an installed capacity of only about 0.1GW, this is a very ambitious goal, but with full support from the government, domestic and foreign investors are interested. Considering 12 GW capacity, it suggests that the role of floating offshore wind farm at depths of 50 m or more is very significant. South Korea has strong domestic manufacturing capabilities, including three turbine manufacturers (Doosan Enerbility, Unison, and Hyosung H&I). At the same time, the government is eager to use Ulsan's shipbuilding industry and offshore plant industry as a new economic opportunity by merging it with the wind power industry to create jobs. In particular, Ulsan is recognized as an ideal hub for floating wind farm due to its excellent shipbuilding/plant industry infrastructure and proximity to deep sea areas, and the local government is also actively promoting the development of floating wind farms. There is also a strong political and public will to increase the share of renewable energy generation to reduce dependence on fossil fuel imports, improve air quality and reduce emissions.

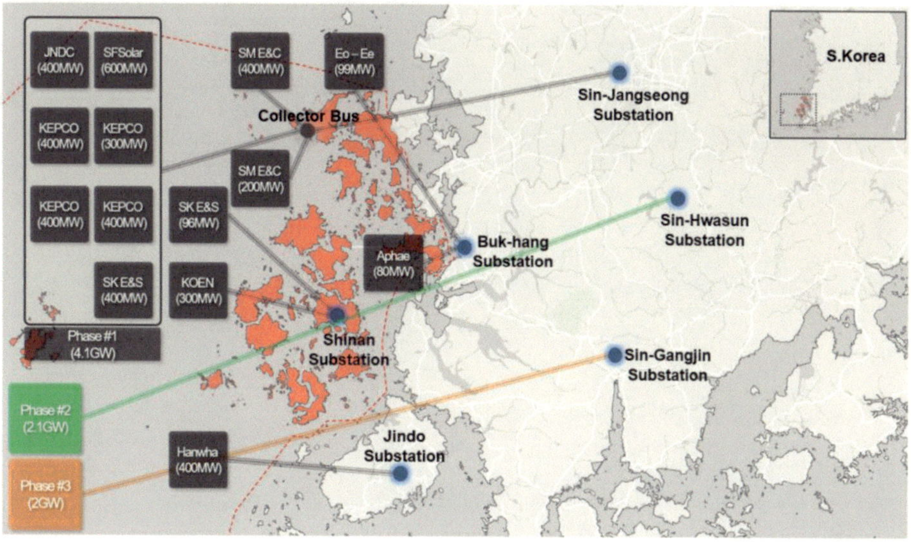

Figure 2.
Shinan 8.2GW project grid plan. The red dotted line and shades of pink mean Shinan region. Black, green, and brown colors represent offshore wind farm project phases 1, 2, and 3, respectively, and blue circles represent substations.

As part of the Ulsan Green New Deal project, Ulsan City announced that it would invest 36 trillion won ($28 billion) by 2030 through collaboration with domestic and foreign companies to create the world's No. 1. Floating offshore wind farm. Ulsan has the advantage that it can be built using existing infrastructure by using the transmission and distribution grids of the Wolseong Saeul, Kori nuclear power plants and Ulsan coal power plants, and using the 'Donghae-1 Gas Field' as an offshore substation. The creation of the Ulsan floating offshore wind farm is expected to diversify large-scale electricity supply and demand such as Mipo Industrial Complex and Ulsan Port, which depended on nuclear power, and create 320,000 jobs. In addition, Ulsan City will lead the global floating offshore wind energy market based on this vision and create an ecosystem for the entire life cycle of green hydrogen production, transportation, storage and utilization utilizing offshore wind power. In addition, Ulsan suggested a direction to take off as a clean energy powerhouse by revitalizing the stagnant local economy by discovering businesses that link sea ranches and marine tourism along with the smooth transition of existing main industries. In particular, Ulsan City decided to use 20% of the electricity produced in the floating offshore wind farm to produce green hydrogen that does not emit carbon. In addition, nine industry-university-research consortiums will be formed to establish a 100 MW class green hydrogen production demonstration facility linked to floating offshore wind power.

Figure 3 and **Table 2** introduce the current status of the floating offshore wind power project in Ulsan. In South Korea, the potential for wind resource is high, especially in the sea near Ulsan, where the water depth is more than 50 m. The average wind speed of this sea area where the projects are being conducted is lower than that of the North Sea, but as a result of floating LiDAR observations, it was confirmed that an average annual wind speed represents 8.8 m/s at 100 m height between 2020 and 2021 [20].

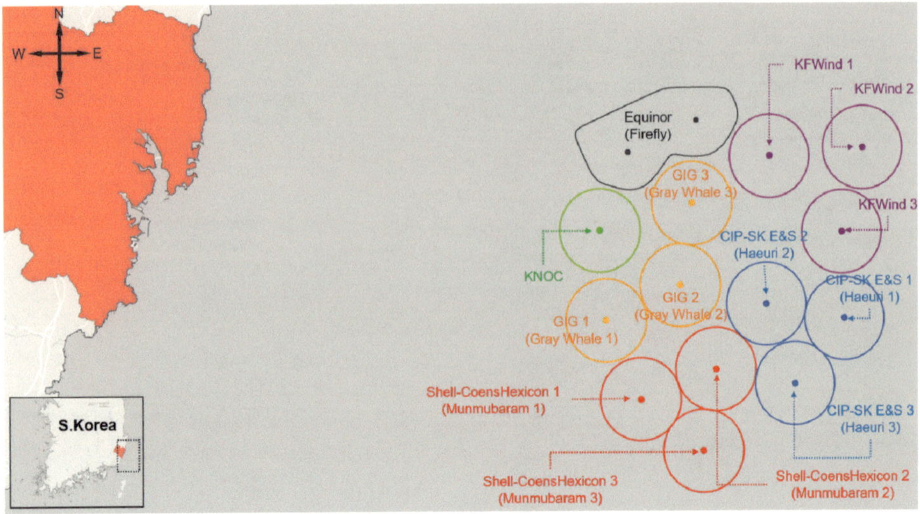

Figure 3.
Ulsan floating offshore wind power project status. Red: Shell-CoensHexicon, blue: CIP-SK E&S, Orange: GIG-Total energies, purple: KFWind, green: KNOC (Korea National oil Corporation), black: Equinor.

Developer	Project Name	Wind Farm Capacity [MW]	EBL
GIG/Total Energies	Gray Whale 1	504	○
	Gray Whale 2	504	○
	Gray Whale 3	504	○
Equinor	Firefly	804	○
Shell/CoensHexicon	Munmubaram 1	420	○
	Munmubaram 2	420	○
	Munmubaram 3	420	○
Korea Floating Wind (KFWind)	KFWind	870	○
	East Blue Power	450	○
CIP	Haeuri 1	520	○
	Haeuri 2	525	×
	Haeuri 3	518	○
KNOC/KOEWP/Equinor	Donghae 1	200	○

KNOC: Korea National Oil Corporation, KOEWP: Korea East–West Power.

Table 2.
Ulsan floating offshore wind power project status.

As of April 2022, Ulsan's floating offshore wind farm project post-licensed capacity is a total of 6.1GW, and all of them received permission from the government on condition of the construction of joint grid access facilities (Collector bus). For the Haeuri 2 project promoted by CIP, the results of the energy business license evaluation are expected in the first half of 2022. RWE and BayWa r.e. have also announced their intention to participate in the Ulsan floating wind project, and are planning to apply for an energy business license starting with a wind resource assessment.

3. Post EBL offshore wind farm projects in the South Korea

In order to conduct a power generation business in South Korea, an energy business license (EBL) must be obtained, and power generation facilities exceeding 3 MW capacity must obtain this license from the Minister of Trade, Industry and Energy. Conditions for obtaining EBL include securing wind resource measurement data for one year at least, WRA (Wind Resource Assessment) reports, consent from residents, and a plan for grid connection to the power system. As of the end of April 2022, a total of 60 offshore wind farm projects (18.13GW) in South Korea have obtained EBLs and are operating or planned (**Figure 4**). More than 70% of these projects are concentrated in Jeonnam and Ulsan. The locations and information of the projects that have obtained the EBLs are shown in **Table 3** and **Figure 5**.

There are 13 projects in Jeonnam that have applied for energy business licenses, but have withheld permission because they do not meet the required conditions. 3 projects in Goheung, 1 project in Wando, 6 projects in Yeosu, 1 project in Jangheung, and 2 projects in Jindo, with a total capacity of about 3.5 GW. Details of these projects cannot be disclosed, but the reasons for the suspension of EBL permits are found to be the failure to adequately solve the problem of resident acceptability, the lack of a grid connection plan, and the overlap with large shrimp farms or

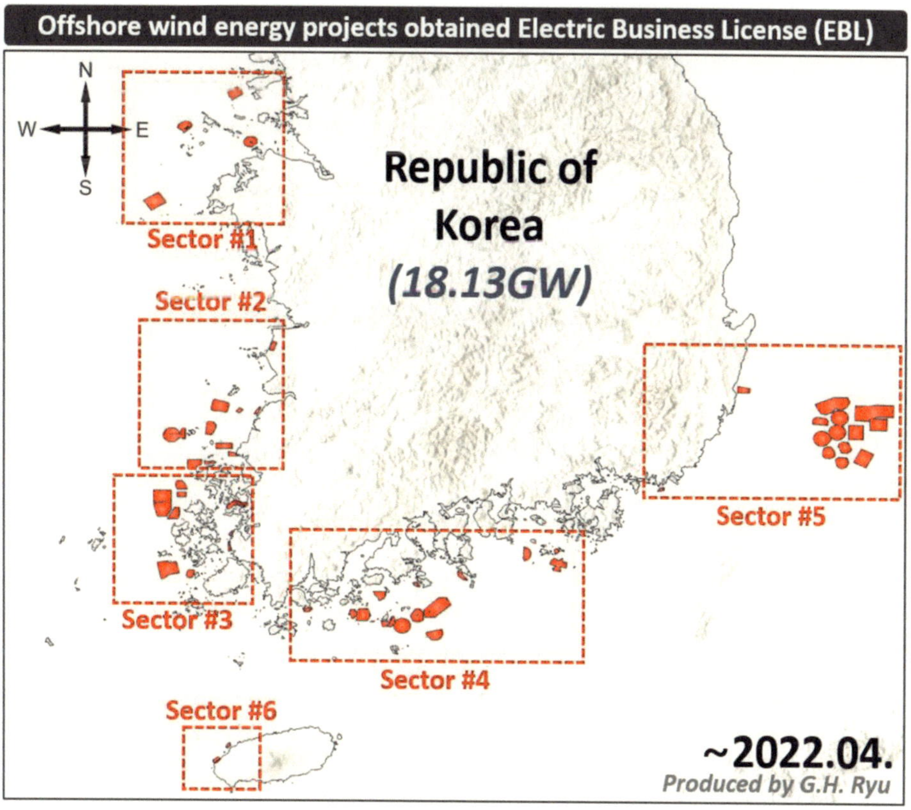

Figure 4.
Current status of offshore wind farm projects obtained for EBLs. 60 projects with more than 18 GW are in operation or planned.

Sector	Capacity [MW]	No. of Projects
#1	1467	5
#2	2655	12
#3	2502	10
#4	4974	16
#5	6406	15
#6	130	2
Total	**18,134**	**60**

Table 3.
Capacity plan of offshore wind farm projects and number of projects by each sector.

military operation zones. The project developers plan to re-apply for EBL by supplementing these issues.

Projects scheduled to start construction in the second half of 2022 are Jeju Hallim offshore wind farm (100 MW), Yeonggwang Nakwol offshore wind farm (354 MW), and Jeonnam offshore wind farm Phase 1 (99 MW). Shinan Ui offshore wind farm (396 MW), Wando Geumil offshore wind power Phases 1 and 2 (600 MW), and Anma offshore wind power Phases 1,2 (528 MW) also appear to be in the final stages of detailed design and wind turbine selection.

Korean offshore wind farm project developers include not only domestic companies such as Hyundai, SK, POSCO, and Hanwha, but also overseas global companies. Ørsted, Equinor, GIG, Total Energies, Shell, CIP, Northland Power, and Vena Energy have entered the Korean offshore wind power market early and are developing their business. In particular, some of these developers are securing competitiveness through joint development by establishing joint venture (JV) with other developers. They are still exploring suitable areas for offshore wind farm to develop additional business plans, and they plan to focus on floating LiDAR by advancing into more distant deep seas.

4. Offshore wind farm support ports and hinterlands

Numerous countries around the world operate offshore wind farms, and additional offshore wind farm projects are underway. This is possible only if there is a port where logistics related offshore wind power are transported and loaded, installation and maintenance vessels can freely enter and depart, and many supply chain companies can coexist [21–24]. In other words, ports are an integral part of the offshore wind farm supply chain, serving as an interface between land-based and offshore activities [22]. In Europe, many countries share the offshore wind farm hub ports of major countries such as Germany and the Netherlands, but Korea cannot create such an environment. The conditions for creating an offshore wind farm support port are shown in **Table 4**, and a study is underway to select offshore wind farm support ports from among the existing ports in Korea that can fully satisfy these conditions.

These offshore wind farm support ports can be classified into three ports according to their roles. The first is a manufacturing port such as turbine manufacturing/logistics management, the second is a construction support port for the installation of an offshore wind farm, and the third is an O&M port for wind farm maintenance.

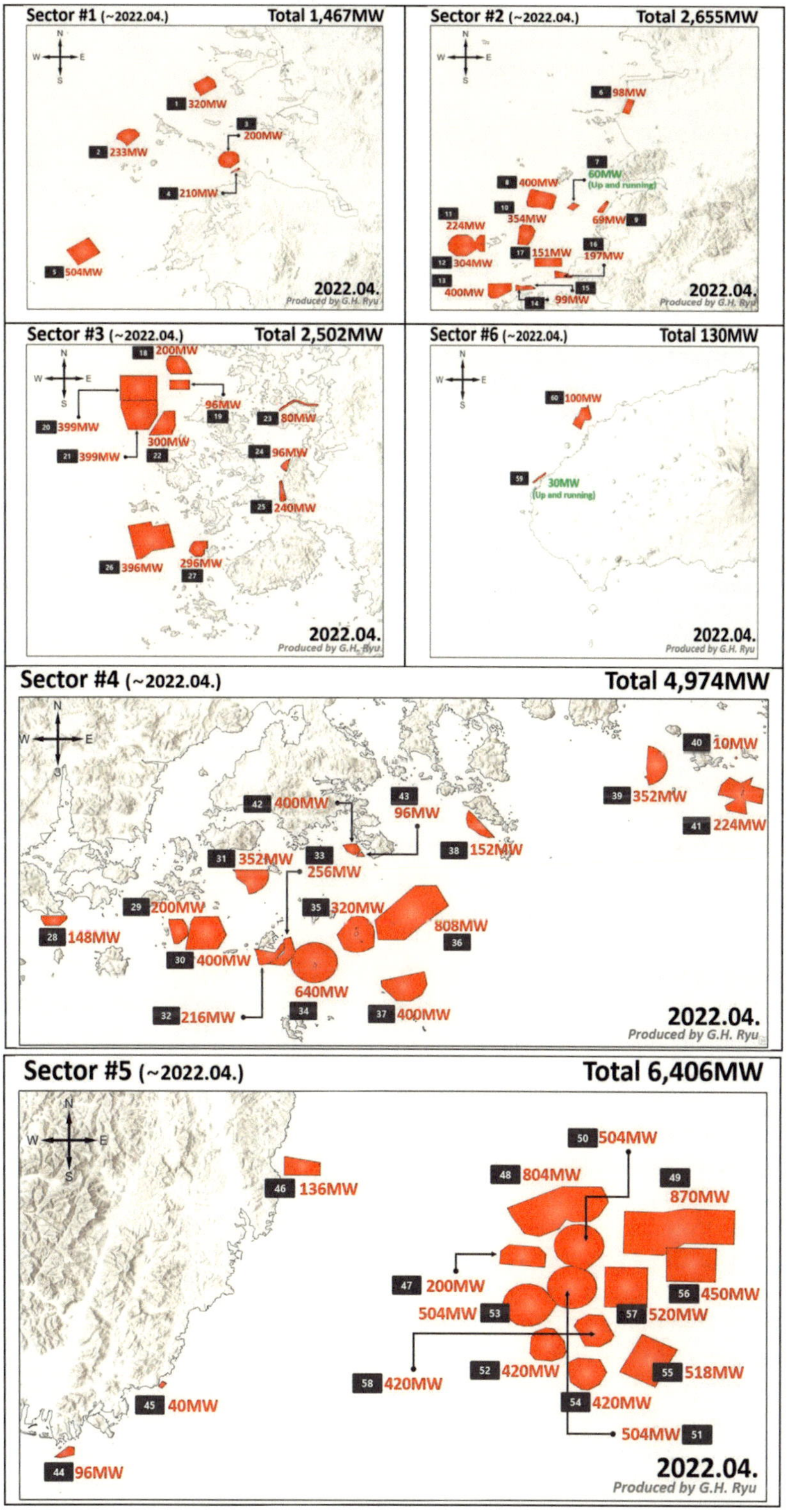

Figure 5.
Detailed site location and planned capacity of EBL-acquired offshore wind farm projects for each sector.

Factor		Contents
Site Condition	Offshore Environment	Water depth
		No. of working days
	Adjacency	Distance between port and OWF
Industry Condition	Integration of Industries	Degree of industrial integration near the port
		Degree of promotion of R&D programs and manpower training
	Technical Skills	Degree of proprietary technology
		Technology level such as parts development
Economic Condition	Transport Volume	OWF projects demand
	Cost (Installation & Operation)	Degree of reduction in business costs
Policy Condition	Government Policies	Whether mid- to long-term related policies are established
		Direct and indirect financial support measures
Social Condition	Civil Complaint	Fishing rights, opposition from neighboring residents

Table 4.
General conditions for development and construction of ports supporting offshore wind farms.

The first manufacturing port is commonly known as a staging or marshaling port, and in some ways it can also be viewed as the final station on a distributed production line that combines secondary components into sub-parts and performs various other finishing operations. Pre-assembly and pre-commissioning of the components (towers and nacelles) to minimize offshore operations are also carried out here.

The construction support port should be relatively close to the offshore wind farm sites in order to shorten the construction period and minimize restrictions from offshore environment changes [25]. This is to reduce the high cost of leasing special wind turbine installation vessels used to transport and install wind turbines and substructures and to maximize working days. In addition, a berth of sufficient length for a large installation vessel to berth should also be provided.

O&M (Operation and Maintenance) ports generally do not require heavy infrastructure due to the small size of the vessels, and generally the requirements are not very different from those of commercial fishing ports. The most important factor in selecting an O&M base port is the distance between the port and the offshore wind farm. It may be more effective if you plan to service several nearby offshore wind farms from one maintenance center.

It is desirable that an offshore wind farm hinterland cluster be formed centered on such a port, and the cluster is preferably composed of a parts production and assembly plant, a manpower training center, an R&D laboratory, a system monitoring center. In South Korea, as shown in **Figure 6**, a total of four ports were primarily selected and their suitability as an offshore wind farm support port is being judged. These are Incheon North Port, Gunsan Port, Mokpo New Port, and Ulsan New Port. As in other countries, the existing large commercial trading ports that managed automobiles, steel, cement, containers, etc. have

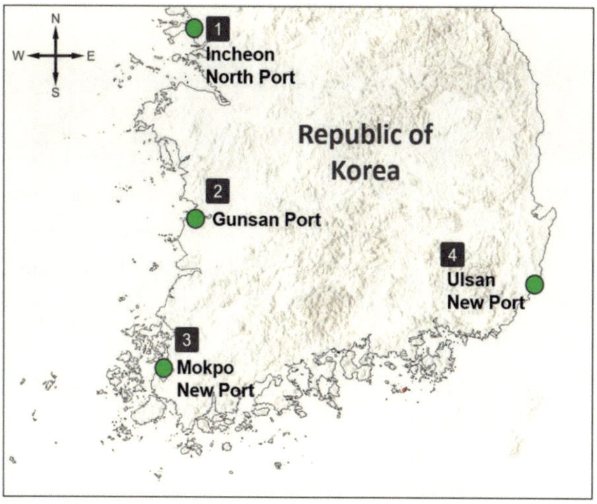

Figure 6.
Commercial ports selected as a target area for an offshore wind farm support port.

many factors that are inappropriate to meet the rapidly changing demands of the offshore wind industry, so detailed analysis and improvement are required. These four ports also intend to change the existing logistics system to form a modified or expanded port.

4.1 Mokpo new port

Since this chapter focuses on offshore wind farm projects in Shinan and Ulsan, It will be taken a closer look at the Mokpo New Port (No. 3) and Ulsan New Port (No. 4). First of all, Mokpo New Port has almost been selected as a support port for the Shinan 8.2GW offshore wind farm, and since it is a port built on a rock rather than a landfill, it has a bearing capacity of several tens of tons [26]. It has the advantage of being close to the offshore wind farms to be built in the Southwest Sea, and as shown in **Figure 7** and **Table 5**, there are many shipyards and industrial estates in the vicinity, so excellent infrastructure has been established.

The Shinan offshore wind farm hinterland complex is being built around the nearby Mokpo New Port and Daeyang Industrial Estate. Mokpo New Port is creating conditions to be in charge of offshore wind turbine/blade/tower manufacturing and logistics and transportation by 2030. As shown in **Figure 8**, the berth of the WTIV (Wind Turbine Installation Vessel), the logistics loading yard, and the parts production line have already been planned. Meanwhile, the Daeyang Industrial Estate plans to build a convergence industrialization platform by attracting and moving in the wind turbine electric/electronic parts manufacturing and offshore wind energy R&D centers. In addition, Daebul Industrial Estate plans to contribute to the creation of offshore wind farms through the manufacturing of offshore wind turbine substructures and machinery through the technological prowess of the existing shipbuilding industrial estate and the attraction of new companies. In addition, the Aphae Industrial Estate of 100,000 m^2 will be built by 2025 to attract wind turbine electric/electronic component manufacturers.

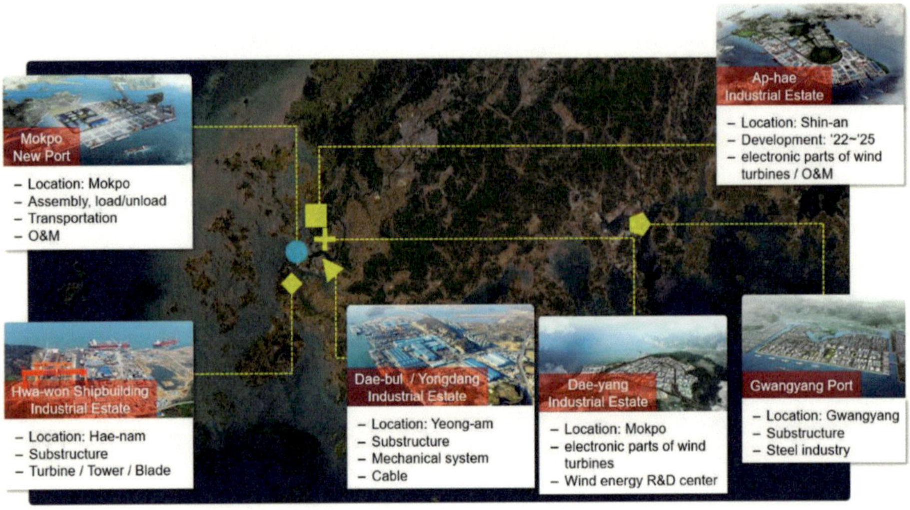

Figure 7.
Current status of offshore wind farm support cluster in the vicinity of Mokpo new port aiming for Shinan 8.2GW offshore wind farm project.

Shape	Name	Type	Location	Role
Circle (Blue)	Mokpo New Port	Port	Mokpo	Support Port (Assembly, Transportation, O&M)
Diamond	Hwa-won Shipbuilding Industrial Estate	Industrial Estate	Haenam	Substructure, Turbine, Tower, Blade
Square	Ap-hae Industrial Estate	Industrial Estate	Shinan	Elcetronic parts of wind turbines, O&M
Cross	Dae-Yang Industrial Estate	Industrial Estate	Mokpo	Elcetronic parts of wind turbines, R&D center
Triangle	Dae-bul /Yongdang Industrial Estate	Industrial Estate	Yeongam	Substructure, Mechanical system, Cable
Pentagon	Gwangyang Port	Port	Gwangyang	Substructure, Steel industry

Table 5.
Composition and role of offshore wind farm support cluster at the hinterland of Mokpo new port.

4.2 Ulsan new port

As a policy, the target of the installed capacity of the Ulsan floating offshore wind farm project by 2030 was initially 6 GW. However, the target has already been achieved in terms of energy business license, and the industry wants to push forward by raising the final target to 9 GW. Ulsan City believes that to build a 9 GW floating offshore wind farm, a dedicated port and hinterland of at least 1 million m^2 is needed. As of the first quarter of 2022, it is in the initial review stage, but a review to expand the port by reclaiming the sea is in progress (**Figure 9**). However, the government does not take the lead, and the method of developing directly by the developer is being focused on. However, unlike Europe, where the shipbuilding industry has stagnated, Korea has an excellent shipbuilding and offshore plant industry and has a steady supply of ships

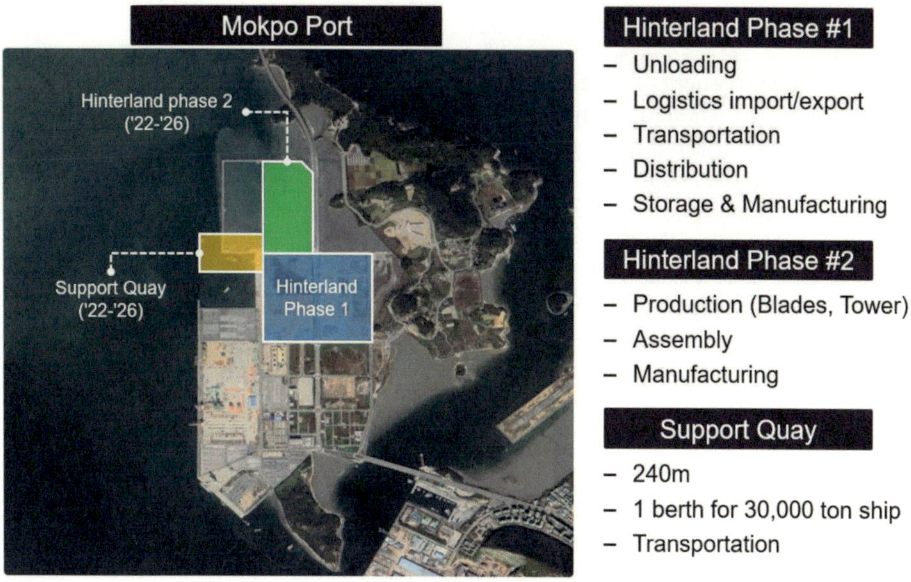

Figure 8.
Mokpo new port offshore wind power wharf development plan.

Figure 9.
Ulsan new port, which is being promoted as an exclusive port for Ulsan floating offshore wind farm projects.

until 2025, making it difficult to renovate and expand it as an offshore wind farm support port. If the private developer gives up the investment, there are many opinions that efforts should be made to reduce the risk as much as possible to the extent that the value of the site that has been built with great effort can plummet. As Ulsan City's efforts are limited, it is suggested that support from the national level is needed. It is necessary to reflect the mid- to long-term possible part in the port development plan so that the floating offshore wind power industry takes root in Ulsan.

5. Conclusions

Offshore wind energy is a main renewable energy source that is being promoted around the world, and is an important means of expanding eco-friendly energy and achieving sustainable energy source supply and demand policies. From the point of view of the national economy, the construction of large-scale offshore wind farms can contribute to economic revitalization and job creation. Also, in the mid- to long-term, it can lead to the revitalization of the offshore wind power industry, and can be a solution that can respond to global climate change.

In this study, Korea's offshore wind power policies, industrial trends, and hinterland ports were briefly introduced. I wrote this chapter in the hope that many foreign project developers and researchers will pay more attention to and invest in Korea's offshore wind power through the contents of this chapter.

Author details

Geon Hwa Ryu[1*], Ji Ye Park[1], Ah Reum Lee[2], Young Gon Kim[1] and Chae Joo Moon[3]

1 Energy Valley Industry-University Convergence Agency (EIUCA), Wind Energy Research Center, Naju, Jeollanam-do, Republic of Korea

2 Enterprise & Advisory, UL Korea, Ltd., Seoul, Republic of Korea

3 Department of Electrical and Control Engineering, Mokpo National University, Mokpo, Jeollanam-do, Republic of Korea

*Address all correspondence to: geonhwa@eiuca.or.kr

References

[1] Ryu GH, Kim Y-G, Kwak SJ, Choi MS, Jeong M-S, Moon C-J. Atmospheric stability effects on offshore and coastal wind resource characteristics in South Korea for developing offshore wind farms. Energies. 2022;**15**(4):1305. DOI: 10.3390/en15041305

[2] Lee J-H, Woo J. Green New Deal policy of South Korea: Policy innovation for a sustainability transition. Sustainability. 2020;**12**(23):10191. DOI: 10.3390/su122310191

[3] Kim J-H, Choi K-R, Yoo S-H. Evaluating the South Korean public perceptions and acceptance of offshore wind farming: Evidence from a choice experiment study. Applied Economics. 2021;**53**(33):3889-3899. DOI: 10.1080/00036846.2021.1888862

[4] Kim C. A review of the development programs, impact, and barriers of renewable energy polices in Korea. Renewable and Sustainable Energy Reviews. 2021;**144**:110870. DOI: 10.1016/j.rser.2021.110870

[5] Kim G, Shin H, Hur J. Probabilistic estimation of wind generating resources based on the spatiotemporal penetration scenarios for power grid expansions. IEEE Access. 2021;**9**:15252-15258. DOI: 10.1109/ACCESS.2021.3052513

[6] Ha K, Kim J-B, Yu Y, Seo H-S. Structural modeling and failure assessment of spar type substructure for 5MW floating offshore wind turbine under extreme conditions in the East Sea. Energies. 2021;**14**(20):6571. DOI: 10.3390/en14206571

[7] Lee K-S, Kim J-H, Yoo S-H. Would people pay a price premium for electricity from domestic wind power facilities? The case of South Korea. Energy Policy. 2021;**156**:112455. DOI: 10.1016/j.enpol.2021.112455

[8] Choi S-Y, Moon B-S, Kim T-G. The maritime environment impact assessment of offshore floating wind power in Ulsan-A focus on habitat equivalence analysis. Journal of Navigation and Port Research. 2021;**45**(3):130-137. DOI: 10.5394/KINPR.2021.45.3.130

[9] Hwang J-W, Kook JK, Jeong H-B, Mun J-H, Kim S. Review of status of wind power generation in South Korea: Policy, market, and industrial trends. International Journal of Engineering Research and Technology. 2020;**13**(12):4943-4952

[10] Chen J, Kim M-H. Review of recent offshore wind turbine research and optimization methodologies in their design. Journal of Marine Science and Engineering. 2022;**10**(1):28. DOI: 10.3390/jmse10010028

[11] Lim J, Kim DK. A study on Korea's remaining GHG emissions allowance and capped-emissions trajectories under the Paris Agreement goal. Journal of Climate Change Research. 2021;**12**(3):255-270. DOI: 10.15531/KSCCR.2021.12.3.255

[12] Park M, Park S, Seong B, Choi Y, Jung S. Current status and prospective of offshore wind power to achieve Korean renewable energy 3020 plan. Journal of Korean Society of Environmental Engineers. 2021;**43**(3):196-205

[13] Kim J-W, Kim Y-K. Induced effects of environmentally friendly generations in Korea. Sustainability. 2021;**13**(8):4404. DOI: 10.3390/su13084404

[14] Park J, Kim B. An analysis of South Korea's energy transition policy with regards to offshore wind power development. Renewable and Sustainable Energy Reviews. 2019;**109**:71-84. DOI: 10.1016/j.rser.2019.04.031

[15] Kim J-H, Nam J, Yoo S-H. Public acceptance of a large-scale offshore wind power project in South Korea. Marine Policy. 2020;**120**:104141. DOI: 10.1016/j.marpol.2020.104141

[16] Nam K, Hwangbo S, Yoo C. A deep learning-based forecasting model for renewable energy scenarios to guide sustainable energy policy: A case study of Korea. Renewable and Sustainable Energy Reviews. 2020;**122**:109725. DOI: 10.1016/j.rser.2020.109725

[17] deCastro M, Salvador S, Gómez-Gesteira M, Costoya X, Carvalho D, Sanz-Larruga FJ, et al. Europe, China and the United States: Three different approaches to the development of offshore wind energy. Renewable and Sustainable Energy Reviews. 2019;**109**:55-70. DOI: 10.1016/j.rser.2019.04.025

[18] Choi N, Kim B, Kim D, Park B, Kim S, Lee B. Grid connection studies for large-scale offshore wind farms considering high penetration of regional renewables. Sustainability. 2022;**14**(2):1015. DOI: 10.3390/su14021015

[19] Ki J, Yun S-J, Kim W-C, Oh S, Ha J, Hwangbo E, et al. Local residents' attitudes about wind farms and associated noise annoyance in South Korea. Energy Policy. 2022;**163**:112847. DOI: 10.1016/j.enpol.2022.112847

[20] Ryu GH. Wind resource characteristics analysis report by floating LiDAR for Ulsan offshore wind farm. Report. 2021. pp. 1-53. 2021-EIUCA-WRA-001-B

[21] Irawan CA, Akbari N, F. Jones D, Menachof D. A combined supply chain optimization model for the installation phase of offshore wind projects. International Journal of Production Research. 2018;**56**(3):1189-1207. DOI: 10.1080/00207543.2017.1403661

[22] Junqueira H, Robaina M, Garrido S, Godina R, Matias JCO. Viability of creating an offshore wind energy cluster: A cast study. Applied Sciences. 2021;**11**(1):308. DOI: 10.3390/app11010308

[23] Ren Z, Verma AS, Li Y, Teuwen JJE, Jiang Z. Offshore wind turbine operations and maintenance: A state-of-the-are review. Renewable and Sustainable Energy Reviews. 2021;**144**:110886. DOI: 10.1016/j.rser.2021.110886

[24] Ahsan D, Pedersen S. The influence of stakeholder groups in operation and maintenance services of offshore wind farms: Lesson from Denmark. Renewable Energy. 2018;**125**:819-828. DOI: 10.1016/j.renene.2017.12.098

[25] Roberto L-A, José MY, José AD-N. Offshore wind installation: Analysing the evidence behind improvements in installation time. Renewable and Sustainable Energy Reviews. 2018;**92**:133-145. DOI: 10.1016/j.rser.2018.04.044

[26] Baek I-H. A study on the development plan of Mokpo port. The Journal of Fisheries and Marine Sciences Education. 2021;**33**(3):702-712. DOI: 10.13000/JFMSE.2021.6.33.3.702

Chapter 11

Flexible Demand for Optimzed Microgrid Design and Cost

Fabien Chidanand Robert

Abstract

Although access to energy has been a major enabler for the development of our civilization, more than a billion people remain without access to electricity. Recent improvements in solar technology offer a unique opportunity to achieve global electrification. However, field studies have reported number of project failures. This chapter is dedicated to project developers, engineers, academicians and policymakers, who wish to contribute to rural development. In the first part, it recalls some important features of a well-thought rural electrification project, while proposing to involve students to liaise with local population. In a second part, it presents an original approach for achieving lowest cost solar microgrid design considering the random nature of solar energy and the users' willingness to be flexible in their consumption. The inconvenience (loss of utility) for users has been modeled and the results suggest that rural users are likely to adapt their consumption to the availability of solar energy to reduce their electricity bill. Such a conclusion is likely to apply to interconnected power grid as renewable energy becomes more and more prominent in the energy mix and energy storage remains a costly challenge.

Keywords: microgrid, design, demand side management, Solar, rural electrification

1. Introduction

Access to energy has been a critical factor for our civilization in the achievement of immense progress in the last century. Sadly, a large fraction of the world's population remains without electricity. Recent improvements in affordable, renewable energy generation technology, offer a unique opportunity to achieve global electrification. In fact, there is a global energy transition towards large-scale integration of renewable energy into power systems [1]. Nevertheless, venturing into deployment of solar technology in rural areas may be challenging.

Although this paper does not address the topic of financing microgrids, it is one of the main barriers to their development, as many rural families have little expendable capital and lack access to credit on one hand and entrepreneurs face challenges due to the small size and risky nature of the projects arising from their remoteness, limited demand and poor consumer base on the other [2]. The first part of this paper is a reminder of few important points presented in the literature and that one must keep in mind while venturing into a rural electrification project: a) defining a clear goal for the community; b) having an inclusive approach to possibly involving students in projects; and c) comparing different technical electrification strategies.

The second part presents a novel technical design approach with a sensitivity analysis depending on users' willingness to be flexible in their consumption. The proposed method is then illustrated through simulations. The results are then analyzed and completed with a model for the loss of utility (inconvenience) that users experience due to their adjustment in electricity consumption.

2. Important points discussed in the literature

2.1 Defining rural electrification goals for the community

The main development goals to be achieved by an electrification project influence the strategy of electrification. The goals must be clearly defined before the study of any technical solution. Depending on the stage of development of a village, development goals may focus more on improving the standard of life in rural families, improving public services or supporting the development of income generating activities. Nevertheless, rural electrification is more important when socioeconomic development takes place simultaneously: the creation of employment, entrepreneurs and the middle class, are crucial to achieve global progress [3]. It is important to evaluate what the most pressing needs of the population are and to consider how different electricity usages can impact development goals [4]. Understanding this relationship will guarantee that the solar electrification effort will produce the expected results. **Table 1** summarizes and classifies the main possible utilization of electricity and their domain of application. For example, in many villages no toilette is available in or near the house and people practice open defecation in the vicinity of their house. Without light, women have thus to restrain themselves to avoid the risk of being abused in the dark [5]. Outdoor light contributes to establishing a general sense of safety. Outdoor light can also keep wild animals away from habitations [6]. Other key usages are solar pumping to provide access to water [7] and income generating activities.

	Light (indoor and outdoor)	Other appliances	Community applications
Economic progress	Increased working hours for shops	Smart irrigation with pumps; Improved productivity with electrical tools	Income generation (e.g. charging station, repair station...)
Social progress	Social/active life at night	Women empowerment; Internet access	Improved education; Access to information with community e-learning and computers
Environmental and health improvements	Better indoor air quality (replaces Kerosene)	Refrigeration for food (conservation)	Health care centre with refrigeration
Livelihood and safety	Better safety at night (women)	Easy and quick access to water for daily use (women) Appliances for cooking	Access to entertainment with computers

Table 1.
Different usages of electricity and their impact on development.

2.2 An inclusive approach

In the literature, there is a plethora of papers about rural electrification using solar energy. Many discuss technical design and methodology. Some point out that human factors play an important role to guarantee a long-term impact of the electrification project [8]. It has been observed that user's involvement has a disproportionate impact on the chances of success of a rural electrification project [9, 10]. Nevertheless, there are several levels of involvement that can be differentiated. Local population can be involved in the definition of the development objectives, in the design of the technical solution; they can be trained on solar photovoltaic (PV) installation and operation practices; finally, they can participate actively in the management of available solar energy and in the payment collection [11, 12]. It is also important to be aware of the gendered aspects of electricity access; not all types of electrification reach and benefit all groups in the same manner [13]. In this context, electrification generates more positive impacts when complementary activities take place (e.g. capacity building, awareness campaigns, access to credit for entrepreneurs, etc.), and when electrification is supported by proper infrastructures and local institutions [14].

It is essential to have a clear vision and values for the project, to have a broad picture of the local situation (social, political and economic), and to identify trusted individuals to be ambassadors of the solar electrification project locally.

2.3 Choice of technical solution

When specific objectives are clearly defined, a technical expert has to translate them into technical requirements. The amount of electricity requested and information regarding geographical and social constraints will inform future decisions regarding the most suitable electrification solution: centralized with or without grid, individually owned systems or a hybrid solution with decentralized solutions. Stakeholders can compare the different viable, technical solutions and write a detailed technical proposal. The proposal must also include implementation, maintenance and monitoring details. The environmental impact of different solutions and technologies has also to be considered, in particular, battery replacement and recycling. The expected reliability of the electrical supply can be a major issue and must be discussed early. Trained manpower may be lacking, yet one must guarantee the regular maintenance of the system. A simple solution may be to train local manpower. However, maintenance of PV systems by users is rarely successful and spare parts and technical assistance should be readily available. Although this issue has been identified more than 20 years ago [15] it remains one of the main challenges for the long-term operation of microgrids today. Fortunately, technology comes to help with a new range of remote control and supervision tools [16].

Extending the main power grid to a village is a simple solution, yet, not always the most reliable and cost-effective. The cost-effectiveness of grid extension can be evaluated by comparing it with the most cost-effective off-grid electrification solution [17, 18]. However, field studies report poor reliability and voltages issues in rural grids [19, 20]. Communication with official utilities is often cited as non-existent, which affects the feeling of trust and ownership of the population. With generation units situated far away from remote villages grid extension also implies a large amount of Transport and Distribution (T&D) losses paid by utility companies. The average T&D loss in developing countries is often very high (37% in Venezuela, 27% in Niger and 19% in India.) (source: World Bank online database). A mixed approach to rural electrification has recently been suggested where renewable energy generation is installed at strategic locations to reduce T&D losses [21], and where villages can also benefit from local battery back-up to improve reliability of supply [22–24]. Though a mixed approach may be the best from the technical and financial standpoint, such an approach requires convincing local utilities about the viability of the project for them to actually perform the grid extension while allowing the village to function in an autonomous mode during power interruptions.

Autonomous solutions require less support from official bodies. They range from a microgrid powered by a single source of energy to solar lanterns, dispatched in each house. Solar lanterns are the cheapest and quickest way of providing light in a village. However, solar home systems require well prepared awareness campaigns to be sustainable and successful [25].

Figure 1 illustrates the main topologies possible. An example of a comparison of different types of solutions is provided in **Table 2**; such a comparison table is project-specific. Solutions were given a grade from 1 to 4 for different criteria, '4' being the best grade and '1' the lowest. This comparison is provided as an example with grades given merely as a general indicator; each project represents a particular situation, and each solution must be evaluated within its context.

Certain solutions can be implemented with alternative current (AC) and direct current (DC). The main advantages of using DC solutions are 1) savings in cost;

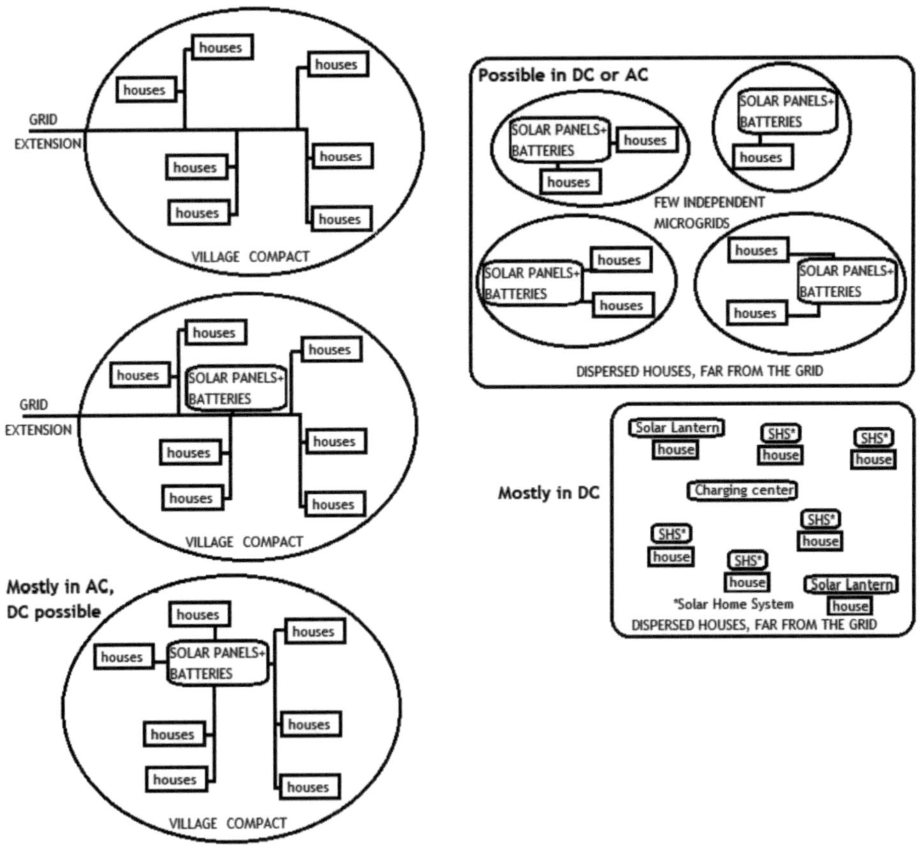

Figure 1.
Different topologies for rural electrification.

	Grid extension	Single microgrid	Grid extension +microgrid	Multiple microgrids	Solar Home System (SHS)	Solar lanterns
Feeling of ownership	1	1–3	2–3	2–4	3–4	2–4
Ease of maintenance	2 (often not reliable)	2	1–2(more complex)	2–3 (Direct Current and modular systems are preferred)	3	3–4 (identical material)
Range of services	4 Can include community services	3	4	3	2 (mostly for individuals)	1
Security of installation	2 (weather)	1 to 3 (theft, and damage)	1 to 3	2 to 3	2 to 4 (misuse)	2 to 4 (misuse)
Social integration	1	2 to 3	2 to 3	2 to 4	2 to 4	3 to 4 (low tech easily assimilated)
Ease implementation	2 to 4	2 to 4 (depends on local support)	1 to 2	2 to 3	3 (rooftop)	4
Affordability	1 to 3 (various distances)	1 to 3 (distance between houses)	2 to 3 (most economical for full service)	2 to 3 (savings on power lines +loss)	1 to 2 (requires more batteries)	4

1 = No/Very little; 4 = Yes/Very much.

Table 2.
Example of comparison of different types of electrification.

2) simplicity; and 3) gain in efficiency. Indeed, DC systems do not require inverters and are most suited for an individual system with small loads. DC is also less dangerous for humans than AC for the same voltage. The main drawback of DC systems is the high losses over long distances and their lack of compatibility with the main grid and standard appliances. However, more and more efficient DC appliances are available.

2.4 The opportunity to involve youth and students

The many years of experience in rural electrification and international development have taught us that fostering the right environment for growth and community development is even more important than having the right technological design. A holistic approach is more complex but allows for a more durable and sustainable impact [26, 27].

In many developing countries, youth constitute a large part of the population in the age of working (in Arica: 37%) but is also largely affected by unemployment (more than 60% of all unemployed people in Africa). It is therefore essential to promote youth leadership and involvement in decision-making to participate in energy access projects [28]. Involving local and international students in such projects has many benefits. Several institutions are now offering their students the opportunity to contribute to rural electrification and more generally, rural development projects.

One of such initiatives is the millennium village project that brings together Columbia University, the United Nations, and industrial partners [29, 30]. Another initiative, on a much larger scale, is proposed in India: Live-in-LabsTM. It was launched by Amrita Vishwa Vidyapeetham, (Amrita University) in 2014 and follows a similar holistic approach aiming at rural development while exposing youth to the lifestyle and problems of the rural population [31]. In Live-in-Labs projects, students get an opportunity to use their skills to contribute to society, thus fulfilling one of the main objectives of education: inculcating ethical values and giving exposure to students. Partnering with an NGO, more than 1000 students, faculty and staff, from Amrita Vishwa Vidyapeetham (Amrita University) have already had the opportunity to serve these villages, living among villagers and experiencing their life. Among the many students' feedbacks, three points are to be highlighted [6]:

- Many students have reported that such projects were a platform for the utilization of their skills, and for experiencing the practical challenges of teamwork.

- There is a new awareness among students about the living conditions of the underprivileged and the challenges they face.

- Most students reported that it was gratifying for them to participate in the design and implementation of a practical solution. Many have stated their wish to continue contributing to similar causes.

The opportunity offered to student to be part of a real project that includes technical and socio-cultural components and associated with a budget and deadlines plays an essential role in preparing them to the active life and completes their education.

3. Methodology for the design of technical solutions

3.1 Overall approach and main parameters

A common approach in the field has been to estimate electricity demand and availability of local resources to design an approximate solution. Nowadays algorithms and software have been developed to optimize the design of autonomous solutions [32, 33]. In general, the design is based on the estimation of a non-flexible electricity requirement that can vary from day to day around an average demand. The load is then modeled with possible time steps ranging from 1 minute to 1 hour. The different technical and economic parameters and weather data that describe the local situation are then added to the simulation software. Possible designs with different combinations of renewable energy sources, diesel generators, grid extension solutions and energy storage possibilities can then be simulated and compared to find the least cost design. Algorithms can perform complex optimizations with multiple objectives [33]; when a professional simulation software is used, the inbuilt algorithm selects the most economical design for a certain level of reliability of supply. In this case, the environmental and social impacts of different solutions can be evaluated and compared manually, by project engineers. **Figure 2** illustrates the general process.

The number of parameters required to perform one simulation or optimization can be excessively high. For example, load can be modeled by a fixed load profile, or by a

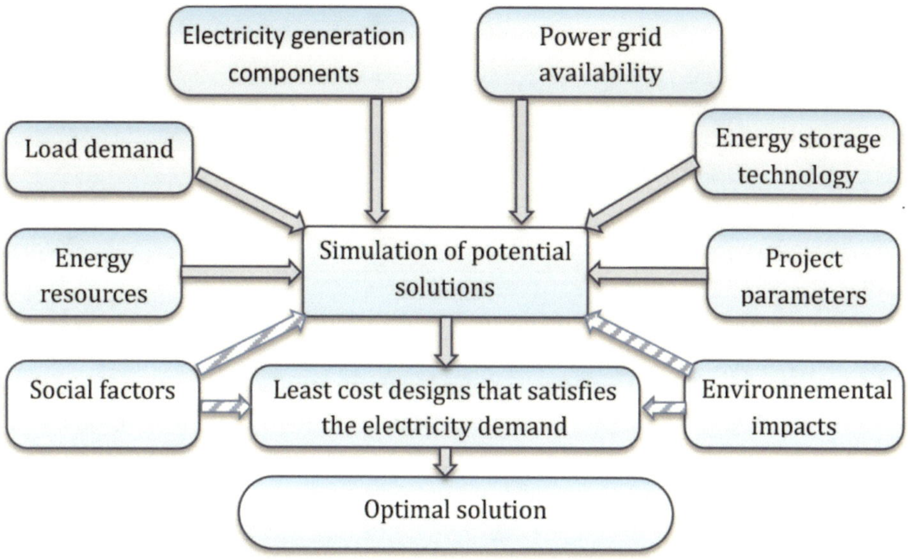

Figure 2.
Rural electrification process to determine the optimal technical solution.

profile with random variation around it. Numerous other parameters are necessary to model energy resources, environmental factors, technical characteristics and costs of energy generation and storage systems, as well as project parameters. A brief summary of the major parameters that influence the design of an autonomous electrification solution is presented in **Table 3**.

Load demand is thus an input parameter among many others that is fed into an algorithm which calculates the 'optimal' combination of technology and energy storage able to satisfy the load at least cost. Design algorithms are compared based on their precision (i.e. their ability to find the best design), computational time and if they can achieve multi-objective optimization. Nevertheless, the precision of the calculated design is limited by the precision of input parameters fed into the algorithm. The two most imprecise parameters are renewable energy resources [23, 34] and load demand. Long-term weather forecast can help optimize the microgrid design while short-term forecast can help better manage the available energy [35, 36]. The accuracy of the load profile immensely influences optimization results [33, 37]. However, with most standalone hybrid energy systems being used in remote and rural areas, load profile data are still scarcely available.

Many field reports highlight that affordability and reliability of supply are critical for the sustainability of isolated solutions used for electrification [8, 38]. Cost of battery storage can be compared to the cost of power outages that can be avoided [39]. Nevertheless, providing high reliability of supply necessitates high investment and thus impacts negatively 'affordability'. On the one hand, low reliability of electrical supply is an obstacle for rural development and on the other hand, affordability is critical for a good portion of rural households. Until now, reliability has been considered to be a firm loss of load. In an isolated microgrid powered by renewable energy, it happens due to a lack of available energy to satisfy firm electricity demand. The proposed method is to introduce flexibility in load demand as a novel parameter for the design of an autonomous solution powered by solar energy. Previous research showed that having sufficient load during daytime reduces the requirement of

Load	Average demand	Load profile	Statistical distribution (variability)	Seasonal variation	Cost of outage	**Flexibility** NEW
Energy resources	Type of source: solar, wind hydro, biomass	Annual Average energy	Distribution of energy throughout the year	Variation around the profile	Environmental parameters: Temperature, air density, humidity,	Derating factors: shading, dust, and surface roughness
Electricity generation components	Power curve, efficiency	Cost (capital, replaceme nt and fuel cost)	Maintenance schedule and cost	Lifetime	Lifecycle environmental impact	Technical settings (site specific): angle of inclination for solar PV, hub height for wind turbine...
Energy storage technology	Max. rate of charge/ discharge and energy content	Efficiency	Lifetime in years and energy throughput Vs. utilization pattern	Environmen tal impact: CO2, metal depletion, and other risk of pollution.	Range of utilization: Depth of Discharge	Condition of operation and derating factors: Temperature coefficient
Project parameters	Project lifetime	Reliability	Inflation	Discount rate	T&D Losses	
Power grid	Distance from specified location	Cost of extension	Grid capacity	Reliability	Environmental impact of electricity produced from the grid	Cost of electricity, and selling price for electricity excess

Table 3.
Major parameters influencing the design of a rural electrical solution.

batteries for an autonomous solution powered by solar (+wind) energy [40–42]. Suitable energy tariffs could be developed to attract such entrepreneurs that consume electricity mostly during the daytime [43]. Once such a microgrid is deployed, shifting load during daytime reduces the probability of loss of load and increases the lifetime of batteries [44].

Mandelli et al. [45], began to detail the different aspects of reliability, energy wastage and their actual costs, when designing a stand-alone microgrid. It is possible to go one step further by evaluating how diverse degrees of flexibility in electricity consumption affect the technical design of an autonomous system powered by solar energy. In other words, the proposed method evaluates the possible savings, on microgrid design, when users agree to adapt their needs to the availability of solar energy (see **Figure 3**).

The flexibility of users was quantified according to three criteria: 1) seasonal: users reduce their consumption during the month with less solar resources; 2) load shifting during daytime: users are invited to shift a portion of their load during sunshine hours; and 3) quick adaptability: users accept that a portion of their needs is adjusted (postponed) according to the real-time availability of solar energy. The savings are evaluated in terms of 1) cost over the project lifetime; 2) reduction in percentage of unused energy (energy in surplus which could not be stored in batteries and was wasted); 3) number of batteries required in the design; and 4) number of solar panels required.

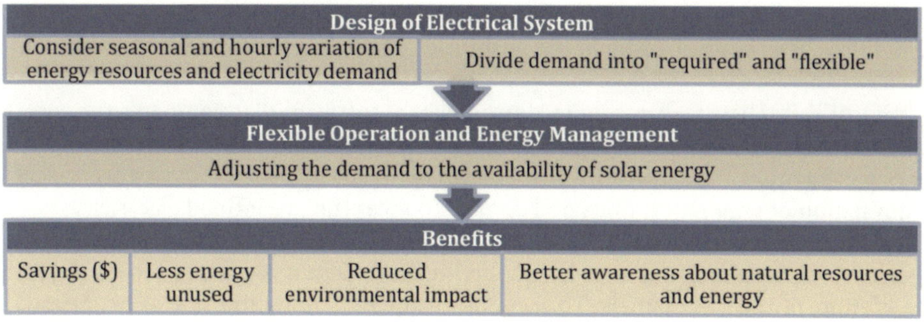

Figure 3.
Proposed approach and its benefits.

The repercussion of the proposed approach, is financial, social and environmental. Such an approach to microgrid design encourages conservation of resource and promotes awareness about the availability of natural resources. It also encourages the development of professional activities that use locally available resources, at the time when they are most available. Here, the notion of degree in flexibility in electricity demand replaces the former notion of reliability. In this context, reliability referred to the occurrence of power interruption due to lack of renewable energy. Flexibility in consumption is rewarded in the form of economic benefits for users. The economic savings comes from two factors: 1) better utilization of renewable energy, with a reduction of unused electricity, previously produced in surplus; and 2) a reduction in the size of energy storage and solar panels. In addition, the environmental footprint of the electricity generation system is minimized.

3.2 Proposed design method and assumptions

3.2.1 Test scenario for the proposed approach

As mentioned earlier, the novel approach is to design an autonomous solution while performing a sensitivity analysis for different degrees of flexibility in electricity demand. A base electricity consumption pattern was assumed to illustrate the method with an average daily load of 50 kWh. The load profile used is shown in **Figure 4**. It corresponds to the profile of a rural settlement of 30 households with 80% of

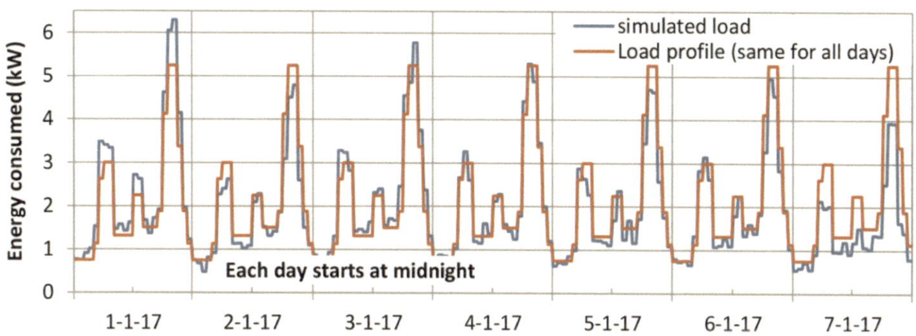

Figure 4.
Chosen load profile and variability over 1 week.

energy consumed by households- mostly in the morning and early at night- and the remaining 20%, consumed by home businesses, enterprises or schools, throughout the day [42].

HOMER Pro, is a well-recognized professional software that compares the cost-effectiveness of different microgrid designs. It can compare different technologies among each other (solar, wind, hydro, biomass.) and suggests the most economical combination of generation units and storage to meet the predefined electricity demand (more details in [46]). The total cost of the microgrid project, called the Net Present Cost, is minimized over its chosen lifetime. Here it is assumed to be 25 years, the lifetime of the solar panels. To account for the remaining lifetime of batteries, a portion of their replacement cost proportional to their remaining lifetime at the end of 25 years, is added to the project. Thus, two designs, one with batteries near their end of life after 25 years, and one with brand new batteries at the end of the 25 years, can be compared fairly.

In HOMER Pro software, loads are represented by a load profile. When a power grid is available, certain loads can be backed-up by the microgrid energy storage while other can be programmed to be interrupted. There are more parameters such as seasonal variation, a difference between weekdays and weekends, flexible loads… The software assumes that the load in each time step varies around the set profile according to a *time step perturbation value* and a *daily perturbation value*. The mechanism for adding day-to-day and time-step-to-time-step variability is simple. First, HOMER assembles the year-long array of load data from the daily profiles specified. Then, in each time step, it multiplies the value in that time step by a perturbation factor α:

$$\alpha = 1 + \delta_d + \delta_{ts} \tag{1}$$

Where, δ_d is the daily perturbation value and δ_{ts} is the time step perturbation value. HOMER Pro randomly draws the daily perturbation value once per day from a normal distribution with a mean of zero and a standard deviation equal to the daily variability input. It randomly draws the time step perturbation value every time step from a normal distribution with a mean of zero and a standard deviation equal to the time-step-to-time-step variability input value. **Figure 4** illustrates the profile chosen for the simulation as well as one-week consumption data with a day-to-day variability and a time-step-to-time-step variability of 10% each, the chosen variability factors for this simulation.

Lead-acid batteries were used for backup as they have the cheapest capital cost though Li-ion batteries are surely becoming an interesting option. The technical characteristics of Lead-acid battery chosen for the simulation are shown in **Table 4**. The lifetime is defined as the maximal length of time or maximum energy throughput for a battery. As soon as one of these values is reached during a simulation, the battery bank

Energy capacity (kWh)	**Cost**			**Lifetime**		**Maximal depth of discharge (%)**
	Capital (€)	**Replacement (€)**	**O&M (€/a)**	**(kWh)**	**(a)**	
1	83	75	1.5	800	8	60

Table 4.
Battery characteristics.

is replaced by the software. The cost of solar panels, inverters, and batteries is the benchmark cost provided by the government of India [47]. The actual solar production of the 31.5 kW solar installation installed on the roof of Amrita University, Amritapuri campus, has been used. The data were measured using a commercial smart meter placed just before the inverter, on the DC cables coming from the solar panels. The software assumes different numbers of solar panels by scaling up or down the measured values. For the microgrid simulations, the inverter and rectifier were both modeled as 95% efficient.

The real-discount rate was assumed to be 7.5% in all simulations [48, 49].

3.2.2 Flexibility criteria

Three flexibility criteria have been introduced and tested. The first parameter tested is the willingness to shift a portion of the load during sunshine hours (8 am to 5 pm). The portion of this flexible load was 0%, 5% or 10%.

The second parameter is seasonal, with a shift of consumption from the month with the least solar energy resource, to other months with maximal solar irradiance. The reduction of load was 0%, 10% or 20% for 1 month, and was compensated by an increased consumption spread over 2 months with higher solar irradiance.

The third parameter proposed is the flexibility to adjust one's consumption throughout the year depending on the availability of solar resources. The portion of load adjusted represents 0%, 5% or 10% of the total electricity consumption.

3.2.3 Test cases

The test cases are summarized in **Table 5**. For a fair comparison, the annual average consumption is the same in all cases. The flexibility of consumption in relation to the availability of the solar resources is thus the only parameter that influences a change in cost for the overall project.

The parameters describing electricity demand and its flexibility have been fed to the software, which calculated the cheapest design able to feed the associated load. A total of eight sets of simulations have been performed. Cases 1A, 2A and 3A are identical, with non-flexible loads (i.e. the software assumes that if a certain microgrid design is not able to feed 100% of the load demand, it is rejected).

	Case N° → ↓	**A**	**B**	**C**
1	Flexible load between 8:00 am and 5:00 pm (% of total load)	0	5%	10%
2	Seasonal variation of demand	0	−10% June and +5% Feb. and March	−20% in June and +10% in Feb. and March
3	Adjusted (reduced) consumption (% of energy shortage in a year)	0	5	10
4	Max. flexibility	Case 1C + 2C + 3C combined		

Table 5.
Test cases with different flexibility criteria and associated parameters.

3.3 Results and analyses

Due to the variable nature of solar resources and electricity consumption, battery storage is required to ensure that electricity is available when users want to consume it. However, batteries are expensive; in the simulation, they represent 70% of the overall project cost in case 1A, and 50% in case 4. Thus, it is sometimes more economical to purchase additional solar panels and let a portion of electricity be wasted (unused) due to a lack of storage capacity. The results of the simulations show the most economical design in all cases. **Figure 5A–D** show respectively, total cost, percentage of electricity unused, (i.e. produced in excess while batteries were full), number of batteries, and number of solar panels. Capital cost is also an important parameter since it often requires the help of donors or subsidies, whereas replacement costs and maintenance costs can more easily be paid by users [2, 8]. **Table 6** illustrates the number of solar panels, number of batteries, total cost and capital cost for the most economical design in each set of simulations.

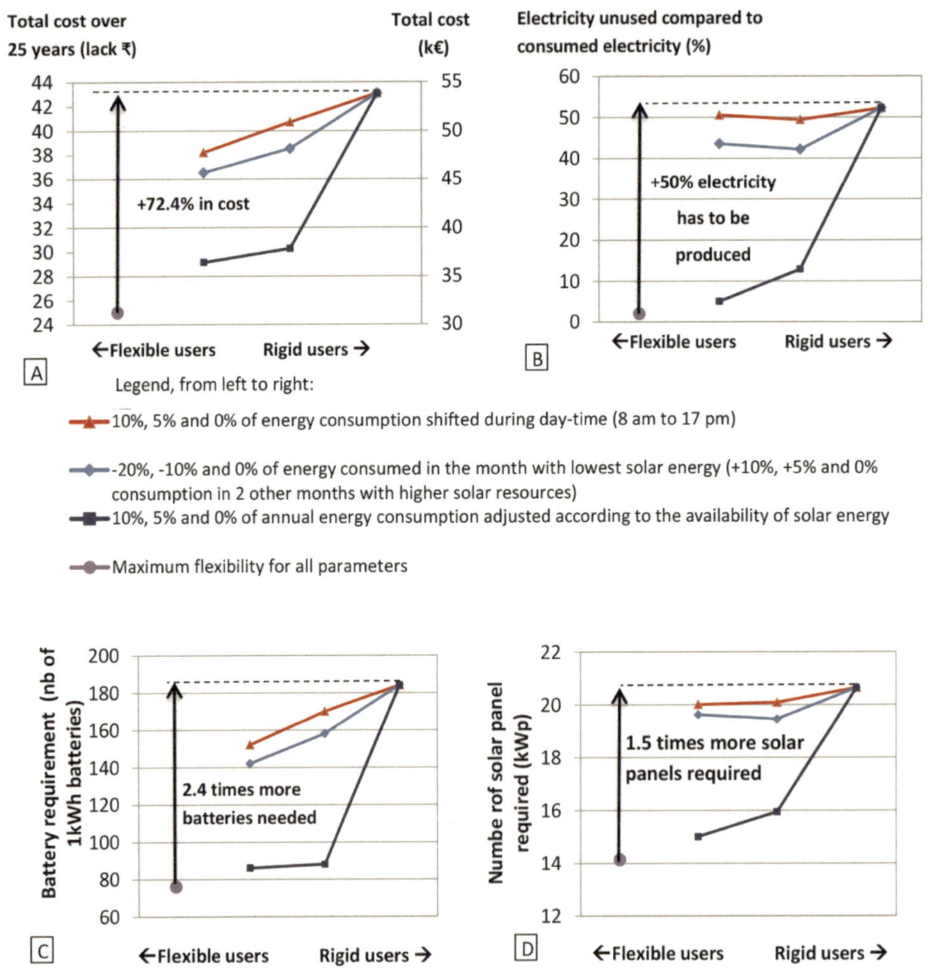

Figure 5.
Microgrid design and total costs depending on users' flexibility. Influence of willingness of users to be flexible on optimized microgrid: total cost (A), unused/surplus electricity (B), KWh of battery (C) and kWp of solar panels (D).

Figure 5 illustrates that the more flexible users are, the less expensive is the overall cost of microgrid.

Summary of key findings

- If users accept to shift 10% of their load during daytime compared to their preferred load pattern, the overall project cost is reduced by 11%.
- If users agree to reduce their load by 20% during the month with the least solar resource (June in this case), the saving on project cost are of 15%, and users have the possibility of consuming 10% electricity more than usual in the two months with the most solar resources.
- If users agree that over a whole year, 10% of their electrical need may not be met, due to lack of solar resources, the savings are of 32% on the overall project cost.
- When users are the most flexible, with 10% of their load shifted during daytime, -20% of load during the month with the least solar resources, and 10% of the remaining load demand to be adjusted as per solar resource availability, the microgrid can be designed at the lowest cost. Comparatively, non flexible users must pay 72% more for their electricity. Indeed, 1.5 times more solar panels are required and 2.5 times more batteries are required for non-flexible users. To meet a none-flexible demand, 50% of energy would be produced in excess of the demand and would remain unused against 2% for fully flexible users.

Shifting load during sunshine hours reduces overall project cost, mostly because it reduces the number of batteries required. When 10% of consumption shifted during daytime, the optimal number of 1kWh battery recommended by the software (to minimize cost), was 153 compared to 184 (21% more) when no load was shifted. When 10% load is shifted during daytime, solar electricity can be directly consumed and the round-trip losses that happen during the storage process are avoided. The savings in round-trip losses also helped reduce the number of solar panels slightly.

Shifting load away from the month with lowest solar resources reduces overall cost because some of the batteries and solar panels are otherwise purchased only to

Case	Solar panels (kWp)	1kWh battery	Total cost		Initial capital	
			(k€)	Savings (%)	(k€)	Savings (%)
1A-2A-3A	20.6	184	53.8	0	33.9	0
1B	20.1	170	50.9	5.4	32.2	5
1C	20.0	152	47.8	11.2	30.6	9.7
2B	19.5	158	48.2	10.4	30.6	9.7
2C	19.6	142	45.7	15.1	29.4	13.3
3B	15.9	88	37.8	29.7	21.7	36
3C	15.0	86	36.4	32.3	20.6	39.2
4	14.1	76	31.2	42	19.0	44

Table 6.
Most economical microgrid design and cost for different flexibility levels of users.

supply the full load during the month with lowest resource (monsoon in India). This equipment may not be useful for the rest of the year, resulting in electricity produced, but unused.

Adjusting ones consumption throughout the year according to the availability of solar resources has the most impact on overall cost. This is due to the unpredictable nature of renewable energy. Indeed, the software has to plan for sufficient solar panels and batteries to supply the full load even when a few days of bad weather occur in a row. When users agree to adjust their consumption based on the availability of solar energy, the design can be optimized. For the same amount of electricity delivered, a microgrid designed to supply an average of 90% of a load of 55kWh requires only 86 batteries and 15kWp solar panels, while a microgrid designed to supply 100% of a load with a daily average of 50kWh, requires 184 batteries and 20.6kWp solar panels. Though both designs supply an average load of 50 kWh/day, in the case where 10% electricity demand is adjusted (shifted), almost all the electricity produced can be consumed. On the opposite, when users are none-flexible, the additional solar panels produce 50% of electricity in excess of the demand, which remains unused.

3.4 Modeling discomfort of users when their demand is not met

Users experience discomfort when part of their demand is not met due to a lack of solar energy resources and a low level of stored energy. This discomfort can be simplistically modeled as the loss of opportunity compared to the utility users would have benefited if their full electricity demand would have been met by the microgrid. In the literature, the benefit that users get from consuming electricity (i.e. the utility) is modeled as a second-order polynomial function [50]. Thus, for the same amount of energy curtailed the distribution of energy shortage in a year can impact the discomfort felt by users. A simple example clarifies this principle in **Figure 6**. The curve illustrates a typical utility function that represents the benefits that users get from consuming different amounts of electricity. The first few kilowatt-hours provide the most benefits, while the marginal utility (i.e. the benefits users get from an additional unit of electricity) reduces and even saturates. At the point of saturation, users do not get any additional benefits from

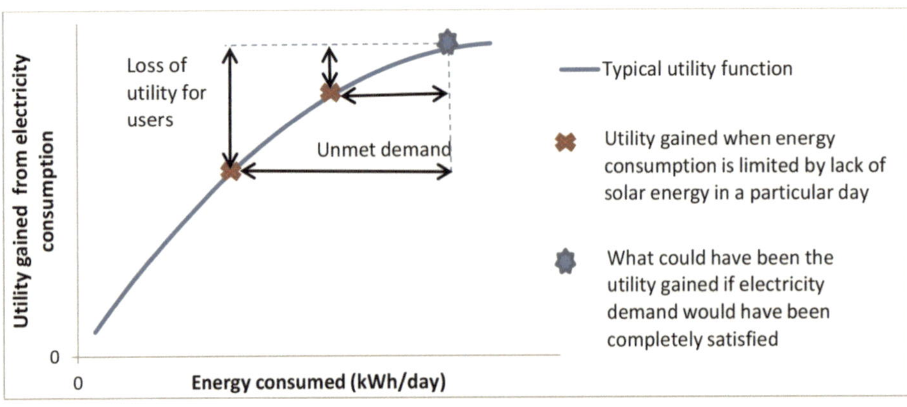

Figure 6.
Example of utility function and loss of utility due to electricity shortage.

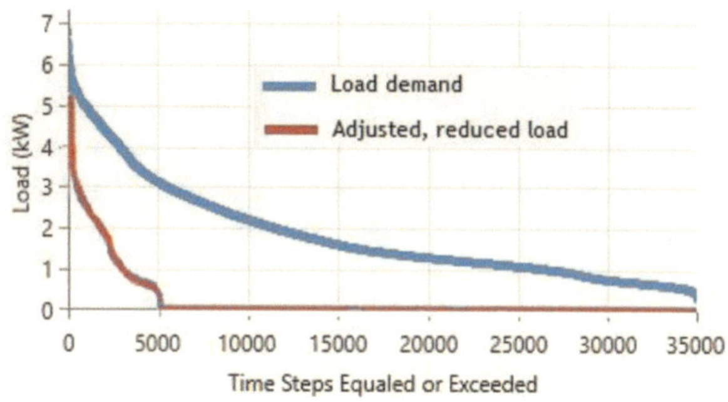

Figure 7.
Satisfied load vs. distribution of adjusted load by time step for an average of 10% adjusted load over a year.

consuming more electricity. From **Figure 6**, one can understand that for the same amount of energy curtailed over a year, users experience less discomfort from having their load curtailed from a few per cent quite often, rather than heavily curtailed but rarely.

A few days of complete black-out may be very detrimental for a workshop or a small business, while limiting the electricity consumption by 30% over a day may lead to a marginal loss. Thus, before agreeing on a certain percentage of flexibility over a year, users must be informed about the expected distribution of energy curtailed. This will allow them to evaluate the loss of comfort or loss of revenue due to adjustments in their electricity consumption. Making the comparison between loss of utility and overall savings on cost of electricity, users can choose the option of flexibility most appropriate to them.

One can evaluate how the electricity curtailment is expected to be distributed over a year. **Figure** 7 shows the incremental distribution of the load curtailed when 10% of the overall demand is unmet (case 3C). The blue curve represents the incremental distribution of the full load demand. The simulation was performed with 15 min time steps so 35,040-time steps in a year. The figure shows that users have to reduce their load 5000-time steps over 35,000 and thus can consume as planned 85% of the time (20 h30/24 h in average). While the peak load observed was above 6 kW, the unmet demand (due to lack of renewable energy) was less than 1 kW for 5% of the time (small reduction in load required for ~1h10min per day on average). The unmet load was between 1 and 3 kW for 10% of the time (fair reduction in load required for ~2 h20 per day on average). The unmet load was above 3 kW for 100 h in a year (important shift of load is required for ~30 min per day on average). With this information, users may prefer to be more or less flexible to spare on cost.

It can be noted that this 'adjusted load distribution' was obtained without considering any energy management algorithm. In practice, users could reduce their load in anticipation of energy shortage to avoid large amounts of energy curtailment. For example, users may reduce their load during a rainy day even though batteries may be nearly full or use simple weather forecasting and thus, avoid power interruption later.

3.5 Main findings and limitations

The sensitivity analysis showed that flexibility of users has a strong impact on microgrid design and should therefore be taken into consideration at the design stage. However, the lack of reliability of the system is also a critical parameter that can lead to user dissatisfaction and non-payment of the electrical bill [8]. It is thus essential that users are aware of the possibility of lack of energy according to the pricing option and microgrid design chosen. When users are informed about the relation between microgrid design, cost, and reliability of supply, they can agree beforehand on a level of cost and flexibility. This approach encourages users to learn about the relationship between their consumption habits, technical equipment, and natural resources. Being aware that consumption habits must follow natural resources availability, can in turn inspire them to care for the system better: avoiding shading and dusting of solar panels. This method of design is also more ecological with less electricity going to waste.

When such an approach is implemented, three additional factors have to be addressed.

- First, users have to develop an ability to predict a shortage of electricity in advance to minimize the impact of energy shortage on their comfort level, or on their business. Thus, an appropriate demand response mechanism is required for users to anticipate energy shortage and reduce their consumption before a power interruption. The literature offers many different mechanisms [49]; an interesting approach consists in settling the demand response process based on a predefined alternative load pattern ranked by preference, 24 hours ahead [51]. Alternatively, a few days of weather forecast, and a simple battery energy level indicator could be broadcasted to all users. Such information will not only improve the usability of renewable energy but also strengthen the feeling of ownership, and responsibility of users towards the energy generating system.

- Second, in a community not all users may accept the same level of flexibility [52]. Poor households may prefer a reduced cost of electricity while entrepreneur, having less flexibility, may prefer a reliable supply, more independent of weather fluctuation. By equaling flexibility to a cost for users, a technical design can be obtained using a single objective/one dimension algorithm and provide a technical solution tailored to the preference of users. Different options are available to satisfy users' preferences. The first option is to design solar home system tailored to each user's requirement. Another option is to prefer small microgrids at different locations in the village, each connected to a group of users with a similar level of flexibility. Research has also shown that it is possible to connect users with a different level of flexibility into a single microgrid [38]. Thus, solutions exist, but they have to be implemented.

- Lastly, electricity demand is evolving over time and predicting its evolution is a real challenge for project with a life-time over 10 years. Existing literature shows that the electricity access-development domain is very complex, dynamic as well as context-and time-specific. Being able to understand and model the aspects and dynamics that determine rural electricity use, can lead to more robust energy planning solutions in rural areas [14].

4. Concluding remarks

This chapter discussed one of the main aspects of renewable energy: its variability, and how it impacts cost and reliability of supply for microgrid users who are not connected to the grid in the context of rural electrification. First, the various electrical usages were classified into categories to gain clarity on how rural electrification can impact the different axes of progress: economic, social, environmental, and related to health, livelihood and safety.. Second, the different rural electrification strategies were classified and rated against key parameters of success to support decision-making by developers. Considering the importance of involving the population early in the project, it was proposed to involve students to liaise more easily with the population; two successful large-scale programs were presented: a win-win approach for the long-term benefit of society: transforming both rural areas but also the students involved.

The full process of technical design was then detailed to achieve the optimal design in diverse scenarios, with key parameters and simulation procedure so far. Instead of considering the load as firm, like in other methodologies, a level of flexibility was introduced as an additional key parameter to achieve a low-cost microgrid design that satisfies users in the context of solar powered microgrid with variable power supply. The proposed method was tested using actual solar production data from the field. The willingness of users to adjust (curtail) 10% of their annual consumption depending on the availability of solar energy, led to 32% savings on the overall project cost. Shifting 10% of load during daytime and lowering consumption by 20% during the month with the lowest solar resources, were found to be effective sources of savings, respectively 11% and 15%. A model was presented to assess the inconvenience for users (loss of utility) to be flexible. This model shows that consenting to a small reduction of load often in the year, is preferred compared to a complete power cut, some of the time. This signals that the new developments in energy management tools such as weather forecast, and user involvement in managing their consumption, can play a decisive role in optimizing comfort of users in a low-cost microgrid powered with solar energy. Their active participation reinforces their feeling of ownership towards the system and transforms simple users into responsible consumers, aware of natural resources availability. On the contrary, an electrification project, aiming at supplying a fixed amount of electricity without considering its final usage and local situation leads to a) under usage of the system and over cost, or b) lack of energy in the microgrid, user dissatisfaction, lower payment recovery, and project failure. In the general context of fast development of renewable energy in the power mix of many countries, it is expected that the same conclusion will apply in interconnected power grids. It is thus expected that in the near future, utilities may offer a wider variety of tariff options to users in order to more accurately reflect the intermittency of power production and the associated cost of energy storage.

To conclude rural electrification: it is more than a technical project; it involves human development and empowerment of individuals. The complexity of interaction between different stakeholders, the various aspects of a project, and the numerous parameters involved makes it hazardous to venture into rural electrification without a proper method. This paper provided guidelines to support project development and academic research towards successful rural electrification using solar energy, with maximum short-term and long-term impacts on the population at the lowest cost.

Acknowledgements

We offer all our gratitude to Mata Amritanandamayi Devi, our chancellor, for supporting and encouraging research that benefits society. We thank David Pinckney for sponsoring the software license and Dr. Vinod Gopal, from the Nano solar department, Amrita Viswha Vidyapeetham, Amritapuri Campus, for providing us with the solar production data.

Author details

Fabien Chidanand Robert[1,2]

1 International Energy Consultants (intec), Germany

2 Amrita Center for Economics and Governance, Amrita Vishwa Vidyapeetham, Amritapuri, India

*Address all correspondence to: chidanand.robert@gmail.com

References

[1] Isabel S, Ferreira P, Lund H. Energy transition: The economics & engineering nexus. Energy. 2019;**166**:961-962

[2] Moner-Girona M, Szabo S, Bhattacharyya S. 1.05 finance mechanisms and incentives for off-grid photovoltaic Technologies in the Solar Belt. In: Comprehensive Renewable Energy. Second ed. Elsevier; 2022. pp. 82-113. ISBN 9780128197349. DOI: 10.1016/B978-0-12-819727-1.00124-2

[3] Zomers A. Remote access: Context, challenges, and obstacles in rural electrification. IEEE Power and Energy Magazine. 2014;**12**(4):26-34

[4] Abhishek J, Saurabh T, Sunil M, Sasmita P, Tauseef S, Karthik G. Access to clean cooking energy and electricity: Survey of states. In: Sen A, Shah M, editors. The Clean Copy, Aspire Design, and Friends Digital. Council on Energy, Environment and Water and Shakti Sustainable Energy Foundation; 2018. Available from: https://www.ceew.in/sites/default/files/CEEW-Access-to-Clean-Cooking-Energy-and-Electricity-11Jan19_0.pdf

[5] Prakash MP, Mohan R, Frey LM, Bonin S. The Impact of Legal Awareness on Enhancing Positive Coping Strategies among SHG Women Subjected to Domestic Violence in a Village, Uttarakhand, India. 5th Int. Conf. On Science and Social Research. 2019. [in press]

[6] Robert FC, Ramanathan U, Mukundan, Durga P, Mohan R. When academia meets rural India: Lessons learnt from a MicroGrid implementation. In: 2016 IEEE Global Humanitarian Technology Conference (GHTC), Seattle, WA: IEEE; 2016. pp. 156-163

[7] Meunier S et al. Influence of the temporal resolution of the water consumption profile on photovoltaic water pumping systems modelling and sizing. In: 2018 7th International Conference on Renewable Energy Research and Applications (ICRERA). Vol. 2018. Paris, France. pp. 494-499. DOI: 10.1109/ICRERA.2018.8566828

[8] Schnitzer D, Lounsbury DS, Carvallo JP, Deshmukh R, Apt J, Kammen DM. Microgrids for Rural Electrification: A Critical Review of Best Practices Based on Seven Case Studies. United Nations Foundation; 2014

[9] Ahlborg H, Sjöstedt M. Small-scale hydropower in Africa: Socio-technical designs for renewable energy in Tanzanian villages. Energy Research & Social Science. 2015;**5**:20-33

[10] Palit D. and Sarangi G.K., 2014. Renewable Energy-Based Rural Electrification: The mini-Grid Experience from India Prepared by The Energy and Resources Institute (TERI) for the Global Network on Energy for Sustainable Development (GNESD).

[11] Hess DJ, Sovacool BK. Sociotechnical matters: Reviewing and integrating science and technology studies with energy social science. Energy Research & Social Science, Volume. 2020, ISSN 2214-6296;**65**. DOI: 10.1016/j.erss.2020.101462

[12] Sovacool BK. A qualitative factor analysis of renewable energy and sustainable energy for all (SE4ALL) in the Asia-Paci_c. Energy Policy. 2013;**59**: 393-403

[13] Winther T, Ulsrud K, Saini A. Solar powered electricity access: Implications

for women's empowerment in rural Kenya. Energy Research & Social Science. 2018;**44**:61-74, ISSN 2214-6296. DOI: 10.1016/j.erss.2018.04.017

[14] Fabio R, Ahlborg H, Hartvigsson E, Pachauri S, Colombo E. Electricity access and rural development: Review of complex socio-economic dynamics and causal diagrams for more appropriate energy modelling. Energy for Sustainable Development. 2018;**43**: 203-223 ISSN 0973-0826

[15] Jafar M. Renewable energy in the South Pacific—Options and constraints. Renewable Energy. 2000;**19**(1–2): 305-309

[16] Philippe J, Beguery P, Barton P. Microgrid Implementation Challenges and Key Technologies. Schneider Electric; 2017. Available from: https://microgridknowledge.com/white-paper/microgrid-implementation-challenges/

[17] Amutha W, Rajini V. Cost benefit and technical analysis of rural electrification alternatives in southern India using HOMER. Renewable and Sustainable Energy Reviews. 2016;**62**: 236-246

[18] Mousavi SA, Zarchi RA, Astaraei FR, Ghasempour R, Khaninezhad FM. Decision-making between renewable energy configurations and grid extension to simultaneously supply electrical power and fresh water in remote villages for five different climate zones. Journal of Cleaner Production. 2021, ISSN 0959-6526;**279**. DOI: 10.1016/j.jclepro.2020.123617

[19] Banerjee SG, Barnes D, Singh B, Mayer K, Samad H. Power for all: Electricity Access Challenge in India. World Bank Studies, Washington, DC: World Bank; 2015

[20] Oyuke A, Penar PH, Howard B. Off-Grid or "off-on": Lack of Access, Unreliable Electricity Supply Still Plague Majority of Africans. Afrobarometer; 2016

[21] Hussain A., Arif S. M., Aslam M. and Shah S. D.A,, 2017. Optimal siting and sizing of tri-generation equipment for developing an autonomous community microgrid considering uncertainties, Sustainable Cities and Society, 32, 318-330.

[22] Palit D, Bandyopadhyay KR. Rural electricity access in South Asia: Is grid extension the remedy? A critical review. Renewable and Sustainable Energy Reviews. 2016;**60**:1505-1515

[23] Robert FC, Gopalan S. Low cost, highly reliable rural electrification through a combination of grid extension and local renewable energy generation. Sustainable Cities and Society. 2018;**42**: 344-354

[24] Samy MM, Mosaad MI, El-Naggar MF, Barakat S. Reliability support of undependable grid using green energy systems: Economic study. IEEE Access. 2021;**9**:14528-14539. DOI: 10.1109/ACCESS.2020.3048487

[25] Venkateswaran J, Solanki CS, Werner K, Yadama GN. Addressing energy poverty in India: A systems perspective on the role of localization, affordability, and saturation in implementing solar technologies. Energy Research & Social Science. 2018;**40**: 205-210

[26] Bhattacharyya SC, Palit D. A critical review of literature on the nexus between central grid and off-grid solutions for expanding access to electricity in sub-Saharan Africa and South Asia. Renewable and Sustainable Energy Reviews. 2021;**141**:110792, ISSN

1364-0321. DOI: 10.1016/j.rser.2021.110792

[27] Robert FC, Frey LM, Sisodia GS. Village development framework through self-help-group entrepreneurship, microcredit, and anchor customers in solar microgrids for cooperative sustainable rural societies. Journal of Rural Studies. 2021;**88**:432-440, ISSN 0743-0167. DOI: 10.1016/j.jrurstud.2021.07.013

[28] Alliance for Rural Electrification (ARE) on behalf of the Africa-EU Energy Partnership (AEEP). Best Practices & Key Recommendations from Young Leaders in Energy Access. Eschborn: Africa-EU Energy Partnership (AEEP); 2018. Available from: https://energypedia.info/wiki/Publication_-_Best_Practices_%26_Key_Recommendations_from_Young_Leaders_in_Energy_Access

[29] Adkins E, Eapen S, Kaluwile F, Nair G, Modi V. Off-grid energy services for the poor: Introducing LED lighting in the millennium villages project in Malawi. Energy Policy. 2010;**38**:1087-1097

[30] Sanchez PA, Palm CA, Sachs JD, Denning GL, Flor R, Harawa J, et al. The African millennium villages. Proceedings of the National Academy of Sciences. 2007;**104**:16775-16780

[31] Ramesh MV, Mohan R, Menon S. Live-in-labs: Rapid translational research and implementation-based program for rural development in India. In: 2016 IEEE Global Humanitarian Technology Conference (GHTC), Seattle, WA. IEEE; 2016. pp. 164-171. DOI: 10.1109/GHTC.2016.7857275

[32] Mahmoud FS, Zaki Diab AA, Ali ZM, El-Sayed A-HM, Alquthami T, Ahmed M, et al. Optimal sizing of smart hybrid renewable energy system using different optimization algorithms. Energy Reports. 2022;**8**:4935-4956, ISSN 2352-4847. DOI: 10.1016/j.egyr.2022.03.197

[33] Al-falahi MDA, Jayasinghe SDG, Enshaei H. A review on recent size optimization methodologies for standalone solar and wind hybrid renewable energy system. Energy Conversion and Management, Volume. 2017;**143**:252-274, ISSN 0196-8904. DOI: 10.1016/j.enconman.2017.04.019

[34] Borujeni MS, Foroud AA, Dideban A. Accurate modeling of uncertainties based on their dynamics analysis in microgrid planning. Solar Energy. 2017;**155**:419-433

[35] Nageem R, Jayabarathi R. Predicting the power output of a grid-connected solar panel using multi-input support vector regression. Procedia Computer Science. 2017;**115**:723-730

[36] Silva DP, Salles JLF, Fardin JF, Pereira MMR, Ottz VC, da Silva FBB, et al. Measured and forecasted weather and power dataset for management of an island and grid-connected microgrid. Data in Brief. 2021;**39** ISSN 2352-3409

[37] Chaurey A, Kandpal TC. Assessment and evaluation of PV based decentralized rural electrification: An overview. Renewable and Sustainable Energy. 2010;**14**:2266-2278

[38] Robert FC, Sisodia SG, Gopalan S. Sustainable trade-off between reliability and electricity prices for geographically isolated communities. Energy Reports. 2019;**5**:1399-1407, ISSN 2352-4847. DOI: 10.1016/j.egyr.2019.09.064

[39] Nayak CK, Nayak MR, Behera R. Simple moving average based capacity optimization for VRLA battery in PV power smoothing application using

MCTLBO. Journal of Energy Storage. 2018;**17**:20-28 ISSN 2352-152X

[40] Bhattacharyya SC. Mini-grid based electrification in Bangladesh: Technical configuration and business analysis. Renewable Energy. 2015;**75**:745-761

[41] Ramchandran N, Pai R, Parihar A. Feasibility assessment of anchor-business-community model for off-grid rural electrification in India. Renewable Energy. 2016;**97**:197-209

[42] Robert F, Sisodia SG, Gopalan S. The critical role of anchor customers in rural microgrids, impact of load factor on energy cost. In: Sixth Inter. Conf. On Computation of Power, Energy, Information and Communication (ICCPEIC 2017). IEEE; 2018

[43] Sisodia GS, Soares I, Banerji S, Van den Poel D. The status of energy Price modelling and its relevance to Marketing in Emerging Economies. Energy Procedia. 2015;**79**:500-505 ISSN 1876-6102

[44] Fezai S, Jamel B. Load profile impact on a stand-alone photovoltaic system. In: 2016 7th International Renewable Energy Congress (IREC). 2016. pp. 1-6

[45] Mandelli S, Brivio C, Colombo E, Merlo M. A sizing methodology based on Levelized cost of supplied and lost energy for off-grid rural electrification systems. Renewable Energy. 2016;**89**: 475-488

[46] Robert FC, Gopalan S. From solar microgrid simulation to field deployment: Accuracy and uncertainties. In: 7th International Conference On Renewable Energy Research And Applications (ICRERA). Vol. 2018. Paris, France; 2018. pp. 1109-1114

[47] Ministry of New and Renewable Energy (MNRE). Benchmark Cost for off-Grid and Decentralized Solar PV Application Programme for the Year 2017–18. Ministry of New and Renewable Energy (MNRE) No.30/11/2012-13/NSM; 2017

[48] IRENA. Renewable Power Generation Costs in 2014. Abu Dhabi: International Renewable Energy Agency; 2015

[49] Robert FC, Sisodia GS, Gopalan S. A critical review on the utilization of storage and demand response on the implementation of renewable energy microgrids. Sustainable Cities and Society. 2018b;**40**:735-745

[50] Samadi P, Mohsenian-Rad AH, Schober R, Wong VWS, Jatskevich J. Optimal Realtime Pricing Algorithm Based on Utility Maximization for Smart Grid. Gaithersburg: First IEEE Inter. Conf. on Smart Grid Communications; 2010. pp. 415-420

[51] Kennedy S, Mnatsakanyan A. A novel demand response model with an application for a virtual power plant. IEEE Transactions on Smart Grid. 2015; **6**:230-237

[52] Nieuwenhout FDJ, van Dijk A, van Dijk VAP, Hirsch D, Lasschuit PE, van Roekel G, et al. Monitoring and Evaluation of Solar Home Systems-Experiences with Applications of Solar PV for Households in Developing Countries. The Netherlands Energy Research Foundation; 2000. ECN Report No. ECN-C-00-089.